THE SKILLS OF PLASTERING

THE SKILLS OF PLASTERING

Mel Baker

MACMILLAN

First published 1990

Published by
MACMILLAN EDUCATION LTD
Houndmills, Basingstoke, Hampshire RG21 2XS
and London
Companies and representatives
throughout the world

Typeset by TecSet Ltd, Wallington, Surrey

Printed in Hong Kong

British Library Cataloguing in Publication Data
Baker, Mel
    The skills of plastering.
    1. Plastering
    I. Title
    693'.6

ISBN 0–333–49981–6

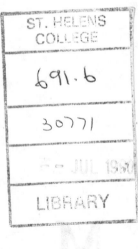

To John Rose — a true craftsman

CONTENTS

ACKNOWLEDGEMENTS

The author and publishers wish to thank the following who have kindly given permission for the use of copyright material.

British Cement Association for material from *Rendering for Farm Buildings,* **45**, 605;

British Gypsum Ltd for material from *The White Book, 1986; Technical Manual of Building Products;* photographic material for various plastering operations;

C & T Metals Ltd for material from a promotional publication;

The Controller of Her Majesty's Stationery Office for material from summary sheets for small contractors;

The photograph of the tool kit on page 1 was taken by Frank Dearling.

Every effort has been made to trace all the copyright holders, but if any have been inadvertently overlooked the publishers will be pleased to make the necessary arrangement at the first opportunity.

INTRODUCTION

This book has been written with the first year trainee or apprentice in mind. It would also be of use to the 'do-it-yourself' enthusiast.

It gives detailed descriptions of working procedures for ceilings, walls and floors, and stresses the need for an awareness of site safety, personal and tool care, and methods of calculating material requirements.

It is intended that a further book will be produced, covering second year craft work.

Plastering can truly be said to be a 'craft' rather than a 'trade'. A plasterer has nothing more than a heap of sand, or a bag of plaster, from which to produce his finished work. The quality of the finished product is dependent, entirely, on the skill of the operative.

The time spent at College will be of great value to the trainee plasterer. Use this time well, for there is much to be learnt.

Do your best to obtain Craft Certificates — they are not just bits of paper, but proof to a potential employer that you have learnt both the practical skills and the technology of your craft. You will, in some cases, have a greater knowledge of craft technology than your employer. If this is so, you will be able to advise and help him, thereby making yourself a valued asset.

New materials are always being developed and literature on these is always available from builders' merchants. Obtain and read such literature, and be prepared to accept changes.

New materials sometimes need different working procedures. Try and adapt to them, and don't condemn them after the first attempt. Carlite plasters became an innovation during my 30 years on site, and I did not like them at first. With a change of attitude and a realisation of their benefits, I would now recommend them at all times. Correctly used, they are no more trouble than traditional sand/cement mixes with a Sirapite finish, and in many cases far more beneficial.

Plastering has come a long way since primitive man 'plastered' the outside of his bamboo hut, and will no doubt develop a lot further yet.

As you develop your skills during your second and third years' training, you may feel a desire to specialise. This could be in the fibrous side of the craft, producing and fitting ornamental plasterwork.

You may, on the other hand, decide that your future lies in the area of machine-applied plaster, which still requires finishing skills.

Wherever you decide your future lies, standards are all-important.

Don't look on your time in College as 'just like being back in school', for it is far more than that. It will provide you with a knowledge of all associated subjects, such as: *industrial studies*, giving an insight into all areas of building development; *calculations*, not sums with no apparent purpose, but methods of working out areas — which will be very important later in your career; and *geometry* for setting out work to arches, circular work, and the design of ornamental plasterwork. All of these will be required if you are to make the most of your acquired skills.

I hope that the book you are about to read will help you along the road to becoming a skilled and respected craftsman, for that is its objective.

CRAFT BACKGROUND

Plastering has a great heritage, dating back to the Greek, Roman and Egyptian civilisations.

All of these races used gypsum, which is the main ingredient of modern plasters. Gypsum was formed, like many other minerals, millions of years ago, and is mined in a similar fashion to coal. The largest gypsum deposit is in the USA. The purest form of gypsum is found near Paris, hence the name plaster of Paris. The colour of plaster depends on the impurities found in the raw gypsum. Shades of either pink or grey are to be found in most plasters. Plaster of Paris, which has fewer impurities, is white.

After the collapse of the Greek, Roman and Egyptian civilisations, gypsum was not used for hundreds of years. During the Middle Ages, houses were built using materials that were readily available, such as stone, timber and straw. Timber-framed houses had internal walls that were made from sheep hurdles. These were 'plastered' with a mixture of mud and straw or with cow dung. This was known as 'wattle and daub'. This form of construction continued until the Fire of London in 1666. Because most of the materials used for building were highly flammable, the rapid spread of fire was inevitable.

From this time, buildings were more substantial and there was a greater use of brick, roof tiles and eventually gypsum. 'Wattle and daub' was replaced by lath and plaster, on a wooden framework. Although this was potentially flammable, it was now covered with at least three coats of sand and lime, which gave protection against fire.

Society was also becoming more affluent, and many rich landowners were having mansions built. Many of these men would have been on tours of Europe, and seen the ornamental plasterwork being produced. This was something previously unknown in England and there were no plasterers capable of carrying out this type of work. This was also the case in Ireland, and it was to Dublin that some Italian craftsmen were first imported. They, in turn, helped to train Irish plasterers, and later both Italian and Irish craftsmen came to England to repeat this training process. The use of decorative plasterwork continued, to a lesser extent, right up to the Second World War.

After the War, with many cities destroyed, there was a need for a rapid re-building programme. Because of this, there was no time available to include time-consuming, decorative work.

It is only now that people are beginning to appreciate, and can afford, decorative plasterwork again. Proof of this can be seen in the number of companies now specialising in this type of work.

1
TOOLS AND EQUIPMENT

As this book is written with the first year apprentice/trainee in mind, I will only deal with the tools that I feel will be required during the initial training period. The plasterer's tool kit is quite extensive, but many of the tools in it will not be required until your second or third year's training.

HAWK

The traditional plasterer's hawk was made of wood, and because it absorbed water, tended to become heavy. At best it was always somewhat cumbersome.

Modern materials now allow the choice of two lightweight hawks — plastic or aluminium. The aluminium type has a detachable handle and the plastic type is a moulded one-piece tool.

In your early days at work, you will, no doubt, find that the pressure of the hawk upon the hand is causing soreness. This can be eased by cutting a hole in a bath sponge, and threading it onto the handle. The sponge will absorb the pressure, and make the hawk more comfortable to use. Most hawks when purchased, come with a rubber pressure ring, but I have yet to find such a ring satisfactory, as it tends to be too thin.

SAFETY

The importance of site safety cannot be stressed too strongly. Never take chances, particularly when working on scaffolds. Always consider both your own safety and the safety of others. Wear a hard hat, a dust mask and protective clothing whenever they are required. Steel toe-capped boots should always be worn.

One of the most common sources of accidents on building sites are pieces of wood left lying around with nails sticking out. These nails can easily penetrate the sole of a boot and cause quite severe injury. Always remove nails from wood before scrapping it. All building sites should have a First Aid box, but I have found that such a box is rarely available when needed. Therefore I suggest that you purchase your own small First Aid box, and always keep it with you. Be sure that it contains an eye bath, for this is perhaps one of the items most required by young plasterers. Make sure, of course, that you also purchase a suitable eye solution.

Take particular care when working around existing electrical installations, and before doing so remove the appropriate fuses from the meter box. **Remember, water or wet plaster and electricity are a potentially lethal combination.**

Sometimes, when using a mixer, your shovel may accidentally get caught inside the moving drum. If this happens, let go of the shovel immediately. Switch off the mixer and then retrieve your shovel.

TROWELS

As your finances permit, you should equip yourself with two trowels — one for floating and one for finishing work. The *finishing trowel* should be of good-quality steel and thinner than the floating trowel. It usually has ten rivets which reach almost both ends of the trowel.

The *floating trowel* is normally not of such good quality, but is more robust in design, having a maximum of seven rivets and a blade of thicker steel.

When you purchase these trowels, do not assume that they must be good because they are new. Faulty rivets can be present, and should be looked for. This is best done by turning the trowel on its side, and looking down the length of the blade. If the trowel has been badly riveted, this will show up as a 'wobble' in the line of the blade. Such a trowel will never produce good work.

FLOATS

You will initially need two floats – a devil (or scratch float) and a skimming float. The *devil float* is used for rubbing-in floating coats, and at the same time providing a key for the finishing plaster. This keying is achieved by tacking small nails into the float (I think doing this at both ends of the float is preferable), allowing the points of the nails to protrude slightly beneath the face of the float. Do not allow the nails to scratch the floating coat too deeply, for this will cause problems when you apply your finishing coat. Allow 2 mm maximum for scratch marks.

The *skimming float* is used for applying finishing plaster to a floated wall. When applying finishing plaster in three-coat work, the skimming float will be used for the second coat. This will be smaller than the devil float.

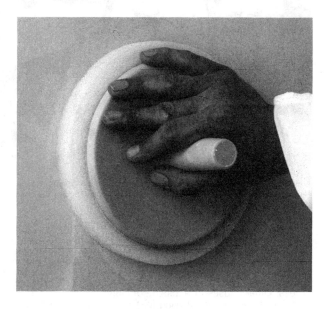

As with modern hawks, floats can now be purchased in plastic one-piece form, as opposed to the traditional wooden float with a separate handle. Wooden floats have a tendency to warp in hot weather, and unless kept in water, when not in use, could distort to the point of splitting. You will have no such problems with a plastic float, for it retains its shape under all conditions and is light to handle.

⬠ Health and Safety Executive   **Construction summary sheet**

SAFE USE OF LADDERS

A man was painting the external window frames of a block of flats from an unsecured ladder. He overreached, the ladder slipped sideways and he fell to his death. The ladder should have been secured, but it would have been even safer to use a scaffold.

Every year many people are killed or injured while using ladders on construction sites. More than half the accidents occur because ladders are not securely placed and fixed, and of these many happen when the work is of 30 minutes' duration or less. Other causes of accidents include climbing ladders while carrying loads, overreaching and overbalancing. This indicates that ladders are being used when other equipment would be safer.

USE OF SCAFFOLDS

There is a temptation to use a ladder for all sorts of work without considering whether the risk involved calls for a better method. It is much safer to work from a properly erected mobile scaffold tower, for instance, than from a ladder.

Jobs such as the removal of cast iron guttering, extensive high level painting, demolition work, or any work which cannot be comfortably reached from a ladder should usually be carried out from scaffolds instead.

SECURING THE LADDER

The foot of the ladder should be supported on a firm level surface and should not rest either on loose material or on other equipment to gain extra height.

Wherever practicable the top of the ladder should be securely fixed to the structure so that it cannot slip. You can use lashings, straps or proprietary clips. While lashings etc are being secured the ladder should be 'footed'.

A ladder fitted with a proprietary spreader arm may be acceptable, provided certain conditions are met. The ladder should, for example, have non-slip feet, be based on a firm level surface which is not slippery, and be erected at a 'safe angle' (see below).

If you cannot secure the ladder at the top you should try to secure it at the base using fixed blocks or cleats, sandbags, stakes embedded in the ground etc.

Where it is not practicable to do this a second person should foot the ladder until the user has returned to the bottom. Serious accidents have occurred because the person responsible for footing a ladder has wandered off to do other work.

Footing is not considered effective for ladders longer than 5m.

SAFE USE OF LADDERS

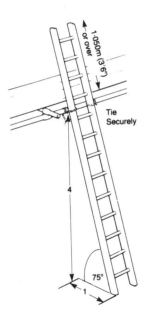

Different grades of ladder are available. Make sure that your ladder is strong enough for the work you do. (The manufacturer or supplier should be able to advise you.)

Avoid overloading ladders - they are liable to break. Only one person should be on the ladder at any one time.

Make sure the ladder is in good condition. Do not use a makeshift ladder and do not carry out makeshift repairs to a damaged ladder.

Have your ladders examined at regular intervals for defects such as cracked stiles and rungs. Don't use defective ladders.

The ladders should extend at least 1.05m above the platform or other landing place or above the highest rung on which the user has to stand, unless there is a suitable handhold to reduce the risk of overbalancing.

Place the ladder at a suitable angle to minimise the risk of it slipping outwards (ideally at about 75° to the horizontal, ie 1m out from the building for every 4m in height).

Rest the top of the ladder against a solid surface. It should not rest against plastic gutters or other such surfaces - appropriate equipment, such as ladder stays, should be used instead. Proprietary spreader arms or similar equipment can be used to span windows or other openings.

LIFTING MATERIALS AND TOOLS

Never try to carry heavy items such as propane cylinders, rolls of felt or long lengths of material up a ladder - you may overbalance, drop the material onto people below or even break the ladder.

Use a small lifting appliance, hoist or rope as appropriate. Carry 'light' tools in a shoulder bag or holster attached to a belt so that both hands are free to hold the ladder.

STEPLADDERS

Stepladders and folding trestles are not designed for any degree of side loading. Workmen have been killed or seriously injured trying to descend from work platforms or landing places using unsecured stepladders.

Do not use the top platform for work (unless it is designed with special handholds) and avoid overreaching. The stepladder is liable to overturn if you do.

SS2 (revised)
Printed and published by the Health and Safety Executive 3/88 100M

GAUGING TROWEL	The gauging trowel is a multi-purpose tool, originally designed for mixing small amounts of material, but can be useful for working behind pipes and other such awkward situations. Do not let your labourer use it as a bucket cleaning tool!
INTERNAL ANGLE TROWEL OR 'TWITCHER'	This is used for finishing-work to internal angles. Twitchers come in two varieties, both of which are equally useful. The most common is the 'box twitch' of three-sided design, but there is also a two-sided version.
HAMMER	The hammer is an essential item, and need not be the traditional plasterer's lath hammer design. A modern steel claw hammer will suffice for everyday needs. Wooden handled hammers will not be as suitable, and probably will not last as long.
SMALL TOOL	A small tool is something you should not be without, and for the purpose of first year employment it will be better if you purchase one of the more robust designs. You will be using this tool mainly for finishing off work in awkward corners, cutting out wet plaster from switch boxes etc. When you become involved in bench work and fibrous plasterwork, you can purchase a small tool more suited to this purpose, for a variety of designs are available.
GP (GENERAL PURPOSE) SAW	This is exactly what the name implies. Worn blades can easily be replaced, because it is a lightweight two-piece tool with a metal blade and a plastic handle.
CUTTING KNIFE	A cutting knife with replaceable blades will be needed for cutting and trimming plasterboards, and will be referred to as a 'board knife'.
TIN SNIPS	Tin snips will be required for cutting metal angle beads and expanded metal and, at a later stage in your training, for other uses.
WATER BRUSH	The water brush has a variety of uses, such as washing off ceiling lines, door and window frames, and wetting down walls, when working with finishing plasters. The best type of brush for general use is a partly worn decorator's emulsion brush. A brand new brush will tend to hold too much water. Worn brushes can sometimes be obtained from your own firm's decorator. If you have to buy one, be sure you inspect it carefully. Some brushes appear to be of good quality, but closer inspection might reveal that the centre of the stock of the brush is devoid of bristles, and is therefore not capable of retaining any water. The minimum width of your water brush should be 150 mm. When not in use, the brush should be washed out in your water bucket and laid to one side, possibly on a window sill. If you leave the brush in the bucket it will be standing on the ends of the bristles and, in time, the bristles will start to curl. This will make the brush virtually useless for cleaning purposes.

BUCKETS

You will need at least two of these, and heavy-duty plastic or rubber ones will last well, provided you look after them. You will always need a bucket of water in your working area. When used for mixing plaster, the buckets must be emptied, scraped out and then washed out. Do not allow plaster to remain in the bucket — it will become hard and difficult to remove. If this should happen, do not beat out the bucket with a hammer, for you will soon puncture it. Because a plastic or rubber bucket is flexible, hard plaster can be removed by gently applying pressure to the outside. The hard plaster will then fall away from the inside.

HAND WHISK OR PLUNGER

This is ideal for mixing plasters in a bucket, and can be a home-made device, although there is a variety of whisks available from builders' merchants and tool shops. These can be used for mixing with either a stirring or a plunging motion. A home-made whisk, for mixing with a stirring motion, can be made with a length of metal tube and some stout wire inserts into the end of the tube. To construct a plunger, you can use a section of stout broom handle and fit a small push-chair wheel to the end of it. Always keep these items clean, and do not lay them down on the floor where they may pick up pieces of grit and dust.

SMALL SHOVEL OR SCOOP

This is ideal for putting plaster into your mixing or 'gauging' bucket, and should not be used for any other purpose.

SHOVEL

A shovel for mixing sand and cement mixtures will normally be provided by your employer or your labourer. The labourer will normally be responsible for maintaining the shovel, but if you do have to use it, make sure you leave it washed off, and clean. Do not beat a shovel with a hammer. If there is a build-up of material on it, then it must be scraped off.

FLOATING AND FEATHER-EDGED RULES

These can now be purchased in a lightweight metal form. However they are likely to be expensive, and you will probably choose to make your own from wood. Ideally, the floating rule should be about 2 metres in length, made of good-quality prepared softwood and about 100 mm × 20 mm. Make sure that the working edge is perfectly straight. The rule should be shaped as shown in the figure. The floating rule is used to straighten the floating coat, and should be kept clean at all times. Always scrape excess material from the rule with a trowel. Do not bang the ends of the rule on the floor. If you do this, the ends of the rule will split and burr over, thereby making the rule useless. Periodically, you should check the rule for straightness to ensure that no warping has occurred.

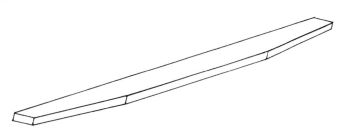

A feather-edge rule is used for checking internal wall and ceiling angles in particular, and its design enables you to work right into the corners and correct any errors. It is generally shorter and lighter than the floating rule.

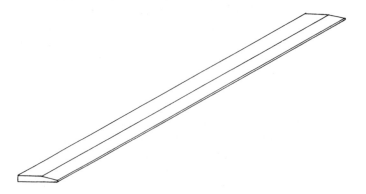

Smaller versions of both floating and feather-edge rules are very useful.

SPOT BOARD
AND STAND

This is an essential piece of equipment for all plastering work, and is sometimes referred to as a board and dummy. These are normally provided by your employer, and are used for placing wet material on, before applying it to the wall or ceiling. You will not require this piece of equipment for floor screeding. The stand is normally made of wood, but metal folding, and rigid stands, can be purchased. The board can be either wooden or metal. The board must be kept clean at all times, and a build-up of old plaster should not be allowed to occur. In effect, the board and stand is a table from which to draw your material.

WHEELBARROW   A wheelbarrow is another essential item, and will be provided by your employer. Always clean the wheelbarrow out when you have finished with it. This can be done by washing it out with the help of a hose-pipe. When removing floating material from the wheelbarrow, you should shovel it from the back with a forward motion. If you work from the front of the wheelbarrow towards the back, it is liable to tip up.

BANKER BOX

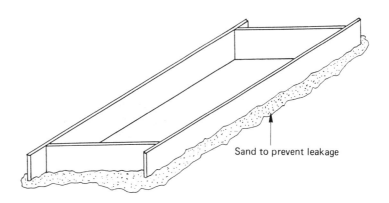

Sand to prevent leakage

This is used for mixing lightweight browning plasters and can be an asset. Browning plasters, because of their nature when in powder form, can be difficult to mix up outside the building. In windy conditions the powder is likely to blow about, and this can also cause eye damage. A good place to mix this type of plaster is in the garage. A banker box can be made from a sheet of 25 mm plywood, or something equally substantial, roughly 2.4 m × 1.2 m in size. This will form the base of your box, but you will need to put sides up to retain the material. Scaffold boards are ideal for this purpose. They can be placed around the base board edges, and held in position with concrete blocks. To prevent material from running out, around the edge of the box, it is a good idea to lay sand around the perimeter to absorb any material that seeps out. This will also help to keep the floor clean. In fact, sand sprinkled around this working area will be beneficial for the same reason. Browning plasters have a setting time of $1\frac{1}{2}$ to 2 hours, so care must be taken to keep the box clean between mixes. Do not lay polythene on the floor, for if this gets wet, as it surely will, it will become slippery and dangerous.

HOP-UP

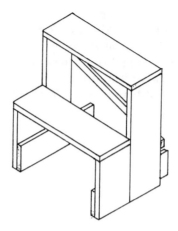

This is the last item of equipment we will cover at this stage, and without doubt one of the most useful. Ideally it should give you two working levels, to suit your own height, and be well made with sound timber.

2
STARTING WORK ON A NEW CONTRACT – SITE PREPARATION

Part of your first day on a new contract will be spent doing preparation work.

If your materials have not been delivered, then there is no need to stand around, waiting for them. There are many jobs that need to be done.

DRY STORAGE
OF MATERIALS

Your first task should be to prepare a storage area for the materials. If there is a garage available, then this is the best place to use, provided that it is dry. Never store materials in damp areas.

A newly laid concrete floor will be damp, so you must lay down a tarpaulin or waterproof material to prevent the dampness getting into the plaster. Having done this, it is a good idea to lay down scaffold boards as well, to stack the materials on.

Plasterboard should be laid flat on the same type of prepared base. When stacked, all materials should be kept covered with a tarpaulin.

The order of stacking materials must be considered. The first thing you are likely to need will be the plasterboards, so you must make sure they are accessible, and not stacked at the back of the garage.

The order of stacking powder materials is also important. If possible, keep different materials in separate stacks. Should this not be possible, then consider in which order you will require them, and stack accordingly.

WATER

You will need to establish that water drums are available. If they are, then clean them out, and fill them up with clean water. This can best be done by using a hose pipe.

In the winter, you must ensure that the hose pipe is drained out each evening, and stored away from frost. The tap must also be protected from frost, likewise your water drum. If you fail to carry out these tasks, and a severe overnight frost occurs, your start on the following day will be delayed while you try and thaw everything out.

BOARD RUNS

Most of the outside preparation work is now done, but you can make access into the building a lot easier if you lay down board runs. This means laying scaffolding boards on the ground, two boards wide minimum, thus forming a 'pavement'. If the ground is uneven, the boards may need supporting with timber to prevent them from sagging or bouncing when you walk along them. Boards can be laid up onto the door step, making entry into the building easy when bringing in your wheelbarrow.

A similar type of board run is useful when unloading materials, particularly in muddy conditions.

INSIDE
PREPARATION

There is now some preparation to be done inside the building.

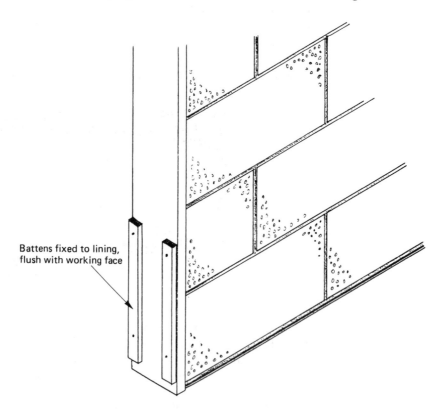

Battens fixed to lining, flush with working face

The door frames and 'linings' will need some protection, so that they are not damaged when you pass through with your wheelbarrow. This can be done by tacking strips of wood to the inside, outer edges of the frames. Pieces of tile batten will be suitable for this, and they can be tacked onto the frame with plasterboard nails. You can now examine each room to make sure that it is ready for you to start work. Check ceiling joists to be sure that you have adequate fixings for your plasterboards, particularly at room edges, where sometimes the spacing of the joists may be such that the plasterboard ends, or edges, cannot be correctly supported.

Check wall surfaces to make sure there are no lumps of hard sand and cement left on them, or nails sticking out of the wall. The best way to examine wall faces is to stand at one end of the wall and look along its length. You will see any obstructions far better than if you look directly at the wall face.

When you are satisfied that everything is in order, and you have carried out any necessary preparation, you can sweep out each room, and dump any rubbish in an appropriate place, outside the building. Do not throw rubbish out of the windows. It can be an advantage to sprinkle sand on the floor to prevent any plaster sticking to it. This can be swept up later.

3
PLASTERBOARD CEILINGS — ERECTING A SCAFFOLD

PRIOR CONSIDERATIONS

Before erecting a working scaffold, it may be advantageous to estimate how many plasterboards you will require for the first room. You can then put the plasterboards into the room, supported upright against a convenient wall in as upright and safe a position as possible. This is the only time that you should stack plasterboards in this manner, for they will be left like this only for a short time. After scaffolding has been erected, the restricted access makes it difficult to take plasterboards into a room without damaging them.

WORKING HEIGHT

You can now erect a scaffold. First you must establish the height of the ceiling and see what is available to give you a comfortable working height. In most domestic dwellings, split-heads can be used. These can be home-made wooden ones or the metal type.

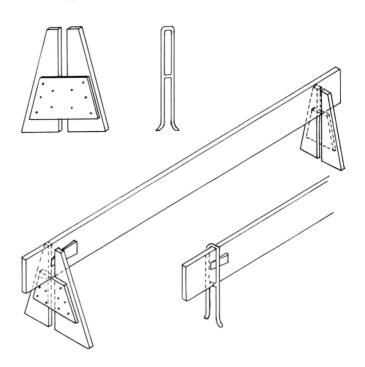

The metal split-heads come in a variety of designs. For medium-height scaffolds, the adjustable-height steel split-heads are suitable. Allow between 150 mm and 200 mm from head height to the ceiling.

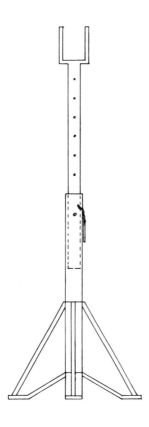

SCAFFOLD BOARDS

You will need to get scaffold boards into the room and you must consider the best way of doing this.

In upper-storey rooms, getting boards up a staircase may be difficult, and there is a risk of damaging the staircase. In this event, it will probably be easier to get the boards into the room by passing them up through an open window. You should not try to do this on your own, for you risk both causing damage and injuring yourself.

If boards are passed in through the window you must be sure to protect the window frame. This can be done by laying thick sacking on the sill, so that the sliding motion of the boards will not damage it. If there is no sacking available, empty plaster or cement bags will suffice, provided you use several bags together, for bags are likely to wear through.

ACCESS TO THE SCAFFOLD

At the point of entry into the room, it is advisable to place a 'hop-up'. This will make your movement on and off the scaffold much easier. You are also less likely to hit your head on the door frame, which can occur if you are looking down to establish your footing.

When constructing the scaffold it is advisable to leave a walking space around the edge, about 300 mm wide. This will allow you to inspect the ceiling from below at a later stage.

PREPARATION OF WALL PLATES

At this stage it is advisable to consider how to prepare the wall plates, in preference to expanded metal lath strips. It is beneficial if wet material can be prevented from coming into contact with wood. Plasterboard strips will achieve this aim.

FIXING PLASTERBOARDS

Do not try and lift full size, 2.4 m × 1.2 m, sheets of plasterboard on your own. Always work in pairs with one man at each end when lifting.

The boards should be fixed with galvanised nails at 150 mm centres. Nails at board edges should not be closer than 13 mm to the board edge. Plasterboard is produced in several thicknesses, so make sure that the nails used will penetrate about 25 mm into the joists.

Where plasterboards are nailed next to each other, a gap of 3 mm should be left between the boards, and at wall edges. Always make allowance for this gap when cutting. Cut edges should, wherever possible, be placed against the wall junction.

APPEARANCE

You will find that one side of a plasterboard is grey (for plastering onto) and the other side is cream (to receive decoration).

There are many different types of plasterboard, but these will be mentioned at a later stage.

4
PLASTERBOARD AND SET TO CEILINGS

TACKING

Having prepared your scaffold in the manner previously stated, you can start plasterboard tacking.

Firstly, mark off, with chalk or pencil, the centres of the ceiling joists, where they meet the wall junction. You must do this, for when you have lifted the plasterboard up to the ceiling, you will be unable to see where the joists are.

Before you fit the first board, you will need to make sure that each end will finish in the centre of a joist. To avoid the tedium of measuring with a tape each time you fit a board, it is a good idea to cut a length of tile batten, or similar light timber, exactly the same length as a plasterboard. This rod can then be offered up to the joists, and all you will need to measure will be the amount to be cut off, should the board ends not finish on a joist.

It is recommended that you use the 'staggered joint method'. This will give extra strength to the ceiling, and make it less prone to movement and cracking on the joints.

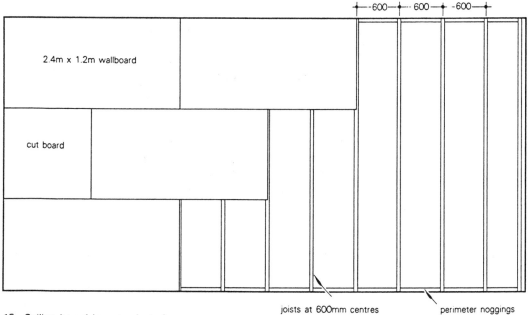

15 Ceiling board layout: plastering
Gyproc wallboard on joists at 600mm centres (maximum)

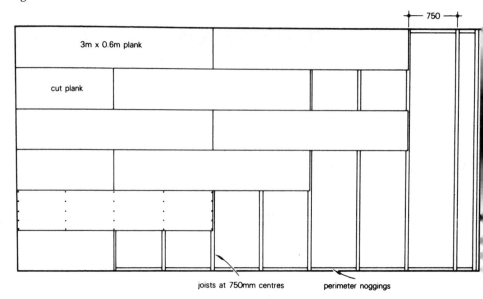

12   Ceiling board layout: plaster or direct decoration
19mm Gyproc plank on joists at 750mm centres
(recommended)

SCRIM

The next task will be to cut scrim for covering all joints between plasterboards. Cotton scrim is popular, but hessian scrim is stronger and more likely to withstand movement. Cut the longer lengths of scrim first, and then the shorter lengths. Do not, under any circumstances, allow the scrims to pass over each other, for this will produce a high spot on the ceiling, which will be difficult to correct. Scrims should be applied to all ceiling/wall joints, and be in contact with both surfaces. As each scrim is cut, it is convenient to tuck it into the joint between the boards where it is to be applied. By this means you will not be confused as to where each scrim belongs.

WETTING THE SPOT BOARD

You can now set up your board and stand in the centre of the scaffold. You will also need a bucket of clean water. Wet the surface of the board all over, using your water brush. This should be done whenever you start plastering, if the board is dry. If this is not done, the wet plaster that comes into contact with a dry board will not be usable, for it will lose moisture and its consistency.

SCRIMMING

Your labourer can now mix the first bucket of Board Finish plaster, making it lump-free and with a creamy consistency.
   At this stage, have only as much plaster mixed as you will need for 'scrimming'. That is to say, applying the scrims to the plasterboard joints. Working on the longest joints first, lay plaster about 150 mm wide, along the joint. While the plaster is still wet, and starting at one end, the scrim can be pressed into the wet plaster. Make sure that the plasterboard joint is in the centre of the scrim. Only work on one scrim length at a time. When the scrim has been pressed into the plaster, lay a little more Board Finish over it and

flatten in. Laying on scrims at ceiling/wall junctions will mean applying plaster to two surfaces, producing different degrees of suction. Therefore, apply the plaster to the low, suction backing first. This will be the plasterboard. Now apply plaster to the wall length, at the junction with the ceiling. The suction from the walling material will be greater, so you must fit your scrim before the plaster starts to dry. If the walling material has a high rate of suction, it is advisable to wet it well, first. Make sure the scrim sits neatly into the angle. Make sure that all scrims are flat and tidy, and not lapped at any point.

PLASTERING THE MAIN AREA

You can begin to plaster the main area of the ceiling, using your finishing trowel. The first coat of plaster should be laid between the scrims and *not* over them, bringing the plaster to the level of the scrims.

For best results, ceilings should be plastered in three coat work, and when the first coat has begun to 'firm-up' slightly, a second coat can be applied with a skimming float. Your trowel will not be needed now, and can be washed. It is best to 'work-in' the ceiling edges first by laying on the material in a strip, as wide as the skimming float length, around the perimeter of the ceiling. Having done this, you can now cover the remaining area. Work away from the ceiling edges, towards the middle.

Now you can check over the ceiling to make sure that it is flat. Pay particular attention to ceiling edges, for if they are not straight it will be very noticeable later. A feather-edge rule should be placed on the ceiling, and if any hollows appear, they should be filled with more material and 'feathered out' with the rule.

When this coat is flattened-in with the skimming float, you can wash off your tools in preparation for applying the final coat. Use your finishing trowel for this coat, and follow the procedure as used for the second coat. You will find that this final coat will use less plaster than the previous coats if you have done your work correctly. While you are waiting for the plaster to become firm, you can wash off your tools and remove the board and stand from the scaffold. The only tools you will need for finishing are a water brush and finishing trowel.

FINISHING

When the plaster appears to start losing its wet appearance, you can test the ceiling to see if it will accept the first trowelling. At this stage you should not need to apply any water to the ceiling. The object of this first trowelling is to flatten-in any ridges. Work as before, edges first and then the main area.

As you proceed with this operation, you should be continually looking for any possible defects. The best way to see any faults is to look toward the part of the room which is the brightest. This will usually mean looking towards the window. If you stand with your back toward the window, you will be looking into the darker area of the room, and will not see things so clearly. Periodically, you should step down from the scaffold and look up at the ceiling, for this is how the ceiling will eventually be seen. Correct any faults that are detected.

When you are satisfied with the quality of the work, the scaffold can be removed, with care, from the room. The scaffold boards can now be moved to the next working area.

CEILINGS ABOVE STAIRCASES

The area of ceiling above the staircase will be difficult to set up. Great care must be taken to ensure that it is absolutely safe. This area of the building is known as the 'well hole', and you will not want to fall into it.

Normal scaffold equipment in tubular form will be erected for you on large contracts. On smaller contracts you will have to arrange this for yourself. This can be done by constructing 'T-piece' sections from stout timber, 150 mm × 50 mm being suitable. You will need two of these on wide staircases, against both side walls at the highest point of the well hole. A further piece of 150 mm × 50 mm must now be accurately cut to fit on the cross-sections of the 'T-pieces', and then nailed together. Scaffold boards can now be laid from the landing area onto the 150 mm × 50 mm cross-section. For added safety, you can lay a double thickness of scaffold boards across this area. Bringing plaster up the stairs will now be difficult, and it may be as well to mix the plaster in an upstairs room. You should arrange this before erecting this scaffold. Alternatively, your labourer may prefer to bring the mixed material up through the window by using a ladder.

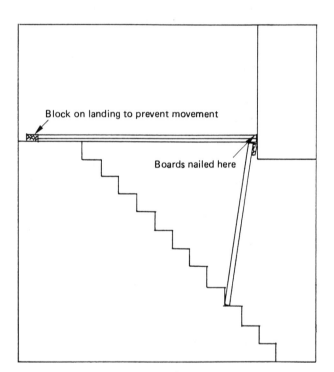

Block on landing to prevent movement

Boards nailed here

REMOVAL OF SCAFFOLD BOARDS

On completion of all ceilings, the scaffold boards must be removed with great care, without causing damage. They should be stacked neatly, outside the building. Do not build up a single stack of boards, for they will tend to wobble and fall over. This will be dangerous. To prevent this, do not stack the boards too high. Form the stack away from the building so that it is not a hindrance or a danger to other workers.

5
PLASTERING TO CONCRETE CEILINGS

Concrete ceilings are not common in most residential housing. They frequently occur in blocks of flats however, and you are likely to have some experience of them in your first year at work. Just as for all other areas of plastering, preparation of the background is the first thing to be considered.

SHUTTERING BOARDS

The construction process of concrete ceilings entails erecting shuttering boards, onto which the concrete is poured. To enable the shuttering boards to be easily removed or 'struck' when the concrete is hard, shuttering oil is applied to the boards. There is a good chance that some of this will remain on the face of the ceiling. To ensure a good plaster adhesion to the concrete, this oil must be removed. Probably the easiest way to do this is to use a bucketful of diluted household detergent. By means of a soft bristled broom, scrub off the ceiling until all traces of oil have been removed. The operation should now be repeated, using clean water. Allow surface moisture to disperse before applying plaster.

RENDER

Carlite Bonding plaster is an ideal product for work to concrete because of its good adhesive qualities, and it can also be used on plasterboard ceilings and walls for additional insulation. Some other plasters will not readily adhere to concrete without more preparation of the concrete, such as applying PVA (Poly Vinyl Acetate) adhesives. Concrete, because it has been vibrated when poured, produces a very dense, low suction background. Even with the use of Carlite Bonding, this will mean that the work should be three coat: render, float and set. Carlite Bonding can be mixed either in a bucket or in the banker box. The banker box is more suitable for large areas. The render coat should be applied first around the ceiling edges then the main area. No straightening will be required at this stage, other than flattening-in. When the work begins to 'steady up', but before it has set, the render coat should be well keyed with a wire scratcher. A temporary scratcher can be made from a piece of EML strip.

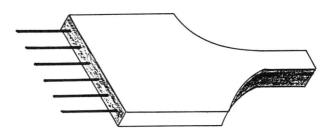

SETTING UP DOTS

To ensure that a perfectly flat ceiling is obtained, the floating coat should be applied with the aid of dots. The setting up of dots can be carried out while waiting for the render coat to set. These dots should be 8 mm in thickness and about 300 mm in length. Set the first dot into a plaster dab at the edge of the ceiling in one corner, and level it in by placing a level on the underside of the dot, along its length. Clean off surplus material from around the edges. Set up the second dot so that it can be reached with the levelling rule, and level it in from the first dot. Check that it is also level in its own length. This process can be repeated around the perimeter of the ceiling, and the last dot checked for level onto the first dot.

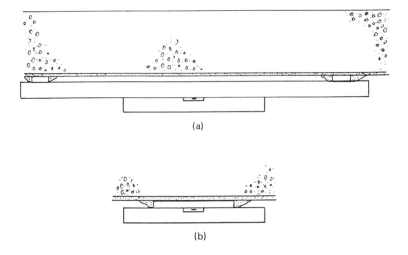

(a)

(b)

The use of a water level and datum line will be covered in your 2nd year work.

On large ceilings it may be necessary to set up dots in the central area of the ceiling. When all this work is completed, you will need to wash off the tools and equipment, and prepare for the floating coat. Keep tools and equipment clean when using Carlite Plasters because of their fast setting time.

FLOATING COAT

As soon as the render coat has set, the floating coat can be applied. Lay on a screed between the dots and flatten in. The feather edge rule is probably best for ruling-off, because of the rather sticky consistency of Carlite Bonding. Start with the feathered edge of the rule in the angle of the ceiling/wall junction and with a side-to-side motion, draw the feather edge towards you.

Fill in any hollows and repeat the process. When all screeds are complete, if the material is firm enough lightly tidy them up with a scratch float, and clean out the internal angles with a floating trowel. Floating can now be applied between the screeds, and ruled off with the feather edge. Remove dots and fill the recesses. Keep the rule fairly flat to the ceiling. If the angle of the rule is too steep, the material will 'tear off'. When the main area of work is firm enough, lightly rub up the work with the scratch float. Take extra care with this on all low suction backgrounds, so that the float does not dig into the work. As the suction on concrete ceilings is very low, it is sometimes advisable to set the ceiling the following day. It is unlikely that the floating coat will have dried out, unless exposed to draught, and it is not likely that any wetting down will be needed.

FINISHING

For best results, three-coat setting will be advisable, using the same procedure as before, with Carlite Finish Plaster. First work around the perimeter, applying the first coat with a finishing trowel, then work in the main area. Apply the next coat with the skimming float, perimeter first, then main area. At this stage, check internal angles with the feather edge rule, and correct as required. When firm enough, rub in the internal angles with the skimming float and clean out the angles with a trowel. As soon as this coat is firm enough, flatten in with either the skimming float or finishing trowel to prepare the ceiling for the final coat.

Wash off all tools and equipment.

Using the same procedure as the first two coats, apply a thin coat of finish plaster. Flatten in the work as it becomes firm enough, with a clean finishing trowel. Clean off all tools and equipment, and remove the board and stand from the scaffold. As the work of finishing the ceilings proceeds, you will need to apply some water to the ceiling to bring up a nice finish. Once the ceiling edges are satisfactory, you may find that the best way of trowelling up is to start at one end of the ceiling and walk along the length of the ceiling, trowelling up as you go. Only work in one direction, checking as you come back to start the next strip. Always work towards the brightest part of the room so that you can more easily see any defects. When the ceiling starts to 'black up', or starts to change colour, you will know that the set of the plaster is almost complete.

Do not over-trowel beyond this point, or the work will become too smooth to receive decorations satisfactorily. As with all ceiling work, you should occasionally get down from the scaffold and inspect the ceiling from floor level.

6
BEAMS — CONCRETE AND PLASTERBOARD

Once having produced a level ceiling, the work of squaring up a beam is made much easier. If the beams are deep, it may be an advantage to work in the sides or cheeks of the beam from the ceiling scaffold, and the bottom or soffit from a scaffold set up at a lower level.

ANGLE BEADS

Angle beads can be used on the external angles. Firstly measure the depth of the beam at several points along its length. Establish the lowest point and, allowing for three-coat work thickness on the soffit, place a large square on the ceiling against the ceiling/beam angle and mark off this point with a pencil mark on the square.

Cut angle beads to length and place dabs of Carlite Bonding at 500 mm intervals along both sides of the angle. Press the bead into the dabs, and place the square against the ceiling, so that the pencil mark is level with the bead. Repeat along the whole length of the bead. You have now established that the bead is parallel to the level ceiling, but you must now check that it is straight along its face length. On a reasonably short beam this can be checked with a straight edge, but on longer beams a line should be held at either end of the beam, pulled tight and the bead adjusted to obtain a straight line. The bead on the other side of the beam should be set in level to, and parallel to, the first bead. When the bead is set in securely the beam should be rendered, floated and set.

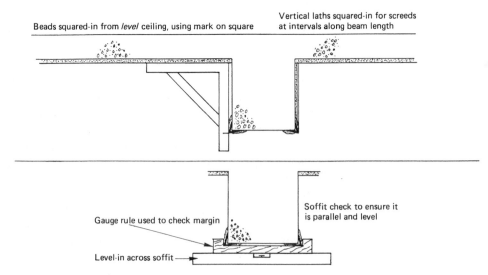

Beads squared-in from *level* ceiling, using mark on square

Vertical laths squared-in for screeds at intervals along beam length

Gauge rule used to check margin

Soffit check to ensure it is parallel and level

Level-in across soffit

24

SETTING UP LEVELS

Working on the sides of the beam will need great care, for they must be perfectly upright and form a 90° junction with the ceiling. This can be achieved by setting-in laths as shown in the figure.

There are more precise methods of setting up levels for ceilings and beams involving the use of datum lines. You will learn these methods during your second year's training.

PLASTERBOARD BEAMS

Much the same procedure will be followed on plasterboard beams except that this will will be carried out in Board Finish plaster. The junction of ceiling and beam will have been scrimmed when the ceiling was set. The external angles of the beam will receive an angle bead, and so the only scrims that need to be applied will be short sections down the sides of the beams and across the soffit on all plasterboard joints. All other work will be the same as for three-coat finishing work to ceilings with Board Finish.

7
FIXING AND PLASTERING ON EXPANDED METAL LATHING (EML)

FIXING EML TO CEILINGS

Expanded metal lathing, apart from being used to cover timber wall plates, is often used as a plastering background in sheet form. Fixing requires great care, whether it is to ceilings or walls.

When used in either of these situations it is likely to be because of the presence of heat, or to withstand the possible presence of heat.

Large ceilings in shops often contain a heating system, above a suspended metal framework. When such a heating system is used, it would obviously not be wise to use a timber ceiling construction. The heat would cause the timbers to warp and this movement would show as cracks in the plastered ceiling beneath. For this reason, EML is the most crack-resistant material suitable for plastering.

Fixing to such a metal framework requires the sheets to be secured with metal, non-corrosive tie wire. This is cut into short lengths and bent double. The loop of wire is pushed through the mesh, over the framework, and back down. When both ends of the wire appear below the ceiling, they are twisted around each other and the ends cut off. When fixing any form of sheet EML it is essential to begin fixing at the centre of the sheet and work towards the ends. The sheet must be kept taut to prevent any buckle occurring. Sheets must be lapped, *not* butt jointed like plasterboards.

FIXING EML TO ANGLES

Fixing around external angles will need some care. Do not try and bend the sheet around an angle by pushing it with your hands, for this will produce a distorted shape.

The most efficient method is to measure the point at which the angle will need to be formed on the sheet, and then place a piece of 100 mm × 50 mm timber flat on the sheet at this point. By kneeling on the timber, draw the sheet upwards and using a second piece of 100 mm × 50 mm place it against the metal and tap along its length with a hammer. This will form a 90° angle in the sheet. This method can also be used when wiring around beams, so that a box shape is formed, thus making fixing easy. By using this method, cutting is kept to a minimum.

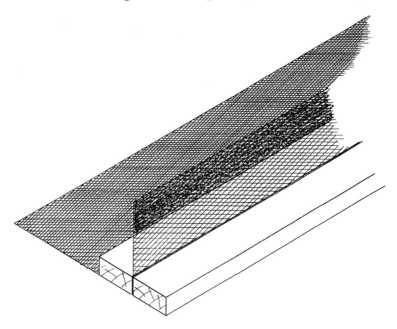

**FIXING EML
TO TIMBER**

Fixing to timber requires the same basic principles. Start at the sheet centre and work towards the ends. Sheets fixed to studwork partitions must be fixed from the bottom of the wall first. As you work up the wall, each sheet must lap over the one below. When fixing to studwork either galvanised nails or galvanised staples must be used.

Ensure that the sheets are fixed taut. Sagging in the EML can cause problems when plastering begins.

**PLASTERING
ON EML**

Modern materials allow you a choice when plastering on EML.

Carlite Metal Lathing plaster is designed for use on EML only.

The first coat applied to the mesh is known as the 'pricking up' coat. When this is applied, less pressure is required than when plastering to solid backgrounds. Too much pressure will result in the material being pushed through the mesh. Lay the material on evenly, and when the set begins to take place, lightly scratch it with a wire scratcher. When pricking up is complete and the material has set, normal floating procedures can begin again using Carlite Metal Lathing plaster to an average thickness of 11 mm from the face of the mesh. Coverage is 60–70 square metres per tonne. Setting time is up to 2 hours. Finishing on Carlite Metal Lathing plaster must be carried out as normal with Carlite Finish. Remember that working on this type of background means that there is little or no suction. Setting the following day is often advisable.

Limelite Browning plaster is also suitable, but a minimum of 24 hours must elapse between pricking up and floating, and 36 hours between floating and setting. Other than this, procedures are the same. Coverage is 120 m^2 per tonne at 11 mm thickness.

8
METAL ANGLE AND STOP BEADS

There are many different designs of angle and stop beads; they help you produce neat angles and stops. With all types of beads, it is only the prepared edges that will show after plastering is completed. At this time, I will only deal with the most common types, and the methods of fixing. All types of beads can be easily cut with tin snips.

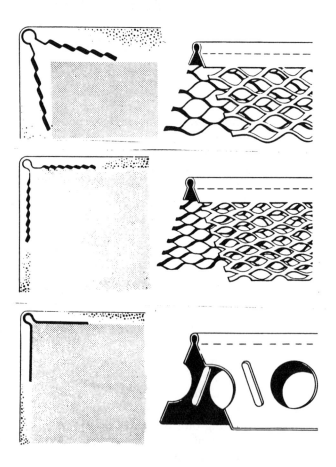

EXTERNAL ANGLE BEADS FOR BRICK OR BLOCK WALLS Beads are available in more than one length, which means you are able to select the length most suited for a particular job. The longest lengths make it possible to avoid joins on long angles. When fitted, angle beads will form a smooth, slightly rounded external angle, and the expanded metal mesh sides of the bead will be buried in the plaster. Angle beads are fitted by applying plaster 'dabs' to both sides of the external angle, about 500 mm apart, and pressing the bead into the wet plaster. A straight edge and level should be placed against both sides of the bead in turn to check for straightness, and to ensure they are perfectly upright. After fitting is complete, clean the bead and wash it off with a water brush. Only work on one bead at a time and, when possible, cut the beads to length before fitting starts. I feel that Board Finish is the best material to use for fixing.

Never fit beads with nails into blockwork or brickwork, for this type of background is likely to be uneven, and the bead will not always fit snugly around the angle. If nailed, the beads will be impossible to move, if repositioning is required, without causing damage. This type of bead will also be fitted over window and door openings where an external angle occurs.

If steel lintels with perforated, recessed faces have been used, then fixing of the bead will be impossible until the recess has been 'dubbed out', or filled with floating material. Whenever possible, this should be done the day before you wish to fit the beads.

When working with Browning plaster, this dubbing out can be done earlier the same day, because Browning plaster has a faster setting time than sand and cement floating material.

Beads for the angles above such openings (the heads) should not be cut until the upright beads have been fitted. Until this time you will be unable to establish the distance between the two uprights.

All beads around window openings should be fitted by squaring off the window frame, making sure that they do not produce an 'out of square' splayed angle.

This is also true with the floor to ceiling beads, which must be squared off the back wall of any recess.

EXTERNAL ANGLE BEADS FOR PLASTERBOARD

Unlike beads for brick and blockwork angles, which are designed to receive a floating and setting coat, beads for plasterboard are designed for Board Finish coats only.

Provided the plasterboard has been fixed cleanly and upright at the external angle, these beads can be tacked on with small galvanised nails.

If this is not the case, then the bead should be fitted with plaster dabs to allow for any correction.

When this type of bead is fitted to an external plasterboard angle, there will be no need to scrim the angle joint.

PLASTER STOP BEADS

These are produced to receive a variety of plaster thickness from thin coat work with Board Finish, up to render, float and set thicknesses.

Situations for their use are very flexible, and they can be adapted easily. They can be used for straight and uniform top edge thickness for cement and sand skirtings, and the same bead type could also be used where a plastered wall ends, and a face brickwork wall begins. With thought and imagination, they can be a most useful item to have at your disposal. Larger sizes can be used at the base of external rendering at damp course level where a 'swept out' effect is required. This is known as a 'bell cast', and has the effect of directing any surface water on the rendering away from the building. All types of bead should be stored carefully. Lay them flat if possible, away from working areas Once damaged they cannot be straightened out.

9
PLASTERING TO WALLS — FLOATING

Plastering to walls will be referred to as either two-coat work or three-coat work.

   Three-coat work means Render, float and set.

   Two-coat work means Float and set.

RENDER

The render coat is not normally required on new work, unless excessive thicknesses have to be made up. Working to old walls however, will very often need a render coat. Usually, this is because the walls are a bad shape, and the low points will need some 'dubbing-out'. All render coats should be scratched well. New work will not need damping down if it has been recently built. However, if the background material has a high rate of suction, it must be wetted down. You can establish the degree of suction in the background by flicking water onto the face of the wall, in a small area. If the water is immediately absorbed, then there is a high suction. If the water trickles down the face of the wall, the suction is low. Old work will need both brushing off and wetting down, to ensure a good adhesion.

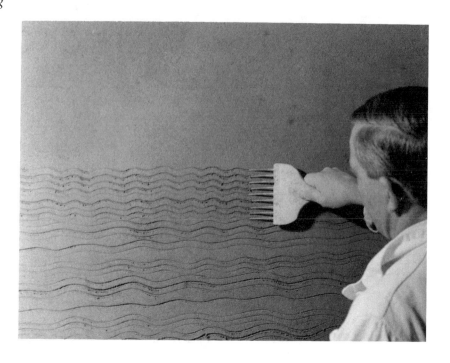

When working up to plastered ceilings, you should wet the ceiling around the edges. Because the ceiling is dry, it will absorb the moisture from the floating material. This will leave an ugly stain on the ceiling, which will be hard to wash off. Wetting down will prevent penetration of the stain.

SCREEDS

The next step will be to apply the screeds. Screeds are horizontal and vertical strips of floating material applied to the wall and straightened with the floating rule. You will rule off from these points when working on the main wall area.

The position of the screeds is arguable. The traditional position for the first horizontal screed is at the head of the door frame height. I feel that this relates to the times when picture rails were commonplace, and it was critical to have a perfect line at this point.

Nowadays, however, I feel it is more beneficial to lay the first screed along at the ceiling junction, thus ensuring a good straight angle at this point. When working on a wall with a fitted door frame, you should find the carpenter has made sure that a straight-edge placed across the face of the frame will give an equal thickness over the whole wall. It should also be perfectly upright. You should always check these points.

Having straightened the top screed, you should now apply vertical screeds to the internal angles. These can be ruled off and, by placing a level on the back edge of the floating rule, they can be 'plumbed up'. When these screeds are satisfactory, you can apply a screed along the base of the wall. This is ruled off as before. This screed is also critical, for the skirting boards will be fitted here, and a bad wall line will be very noticeable.

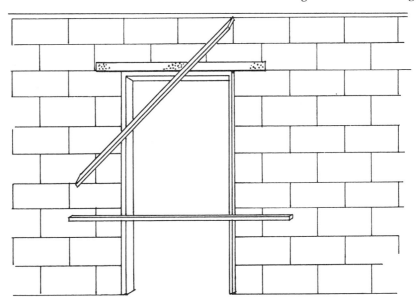

FLOATING TO MAIN AREAS

Now your screeds are complete, you can begin filling in the main wall area. Firstly, offer up the floating rule, with it touching the top and bottom screeds. This will indicate the thickness of floating to be applied, for the wall may not be perfectly straight.

When the material is applied you can begin ruling off, and filling in any hollow points. Your vertical screeds will allow you to cross-check the wall vertically, horizontally and diagonally. When you are satisfied, and the wall is firm enough, it can be rubbed up with the scratch float. Take time with this operation and fill in any holes. Remove any leaves or soil that you may find, for they will stain badly.

CLEANING

Next you must cut out the ceiling angle with your trowel, and wash off the stains with your water brush. Cut out the floating material from any electrical boxes. Cut out cleanly at the base of the wall. Where two walls meet, the internal angle must be cut out with the trowel. Failure to clean out internal angles properly will present problems when you apply the finishing plaster, for by this time the floating coat will be hard.

When all the floating work is complete, you must make sure that all cleaning operations are completed. Cut back slightly around the door linings to allow the finishing plaster to finish flush with the face of the lining. Clean off window frames and sills. Wash off windows. Do not scrape the glass with your trowel for this can cause scratches, and could lead to expensive replacements. Any stubborn lumps of material on glass should be soaked, and then gently eased off with a piece of softwood. Recess slightly up to angle beads, and gently wash off any material. When these operations are complete, the board and stand can be removed from the room and set up ready for the next. The floor must now be cleaned. If any material on the floor is usable, it can be saved. This is quite common when working on timber floors, but material dropped onto concrete floors is not always suitable, since it will have become dry if it has lain on the floor for some time. This can be remedied to some extent by placing 'catch boards' around the base of the wall before you begin floating. Any waste will fall onto the boards, and can be picked up more easily and saved for re-use. Scaffold boards will be suitable for this. When all waste material has been picked up, the floor can be well swept with a stiff bristle broom. The room should be left in such a way that when you come to apply the finishing plaster, little preparation will be required. Scaffold boards used as catch boards should be cleaned off.

BROWNING
PLASTERS

Not all floating work will be carried out with sand and cement mixes. You will quite often be using lightweight Browning plasters. All the operations described for ordinary floating work will be the same. However, Browning plasters have a setting time of about $1\frac{1}{2}$ to 2 hours, so care must be taken to ensure that tools and work areas are kept constantly clean. Working on walls in staircase areas can mean that floating materials will be dropped, making a mess on the staircase. This can be made easier to clean up if catch boards are laid up the sides of the staircase treads. Any material that falls will then land on the boards and slide to the floor below, where it can be easily picked up.

10
PLASTERING TO
WALLS — SETTING

PREPARATORY
WORK

The finishing plaster that is applied to a floated wall is called the setting coat. If a wall has been floated in sand and cement, the finishing coat will be Sirapite. This cannot be applied on the same day as the floating coat. Finishing coats applied to Browning plasters can, however, be applied the same day. This is because the Browning plaster has a fast set, that is to say it will be hard in $1\frac{1}{2}$ to 2 hours. Sand and cement floating coats are best if set the day after floating. By doing this the floating will not have had a chance to dry out too much, thereby making easier working, but at the same time giving a moderate rate of suction. This is referred to as 'setting on a green suction'. If the floating coat has been left too long, or appears to be very dry, you must check the rate of suction by applying water to the wall as described for checking backgrounds. Should the wall need wetting down more than can sensibly be done with a water brush, then water can be thrown onto the wall with a cup or empty tin.

You will find that if you only apply the water to the top half of the wall, the water will find its way to the bottom half. If you throw water onto the bottom half of the wall, it will have a tendency to run off onto the floor. This can be a nuisance, and will not be beneficial to a timber floor. Do not over-wet the floating. You should stop applying water when the floating appears damp all over, but is not holding water on the surface. If you do over-wet the wall, then allow time for the water to be absorbed. Never apply finishing plaster to a dry background. If you do, there will not be a good bond between the two coats, and surface crazing will occur. Application of finishing plaster to lightweight Browning plaster will be carried out the same day as the floating. This will mean that the floating should remain damp and no wetting down will be required. Some thermal blocks have a high rate of suction, and will draw moisture from the floating coat. In this instance, some wetting down will be required.

APPLICATION

When this preparation work is complete, the finishing plaster can be mixed up and placed on the board. The bucket must now be scraped out and washed. Start applying the setting coat by using your hop-up to stand on, and apply a coat of plaster to the top of the wall, along its whole length. This should be brought down low enough to allow the next section of wall to be plastered from a standing position on the floor. The centre section of the wall can now be covered, using upward strokes with the trowel along the wall's length. The same can now be done to the lower section of the wall, working as close to the floor as possible.

Make sure the floor is clean at this point, so that you do not pick up any grit with your trowel. Allow the first coat to 'steady up' before applying the next coat. This means when the 'wet look' disappears from the plaster. You should check this by placing your fingers lightly on the plaster. If no plaster sticks to the finger tips, the wall is ready. For best results, as with ceiling work, the second coat should be applied with a skimming float. Follow the same procedure as for the first coat. Check all internal angles with a feather edge, and correct them as required. Using the float lengthways, rub in all internal angles, and clean them out with your finishing trowel. Check over the main wall area to make sure that no ridges are showing. Lightly rub it over with the skimming float to leave a well prepared face for the final coat. The final coat is applied with the finishing trowel, using the same procedure again. When this has been done, clean off your board and wash off all tools that have been used. Allow the coat to steady up.

FINISHING

When the wall is ready, begin to 'flatten in' the work with the finishing trowel. You should not need to apply too much water to the wall at this time. Possibly no wetting down will be required. You will need to trowel over the whole wall face as many times as required to achieve a good finish. The wall face should not be over-trowelled. A high polish on a wall may look nice, but it will seal the face of the plaster to an extent that will make the application of decorations very difficult. To avoid working-in an internal angle where two areas of wet plaster meet, it is advisable to work on opposite walls first. When these two walls are finished and hard, the other two opposite walls can be set. By doing this, you stand less chance of cutting into the adjoining wall. So that a nice square internal angle is produced, the twitcher can be worked down the angle to fill in any 'misses'.

CLEANING

When all walls have been set, you should start work to leave the room in a clean condition. While the plaster is still wet, clean out electrical power and switch boxes. Wash off any plaster splashes from the windows. Scrape off and wash off window frames and sills. The floor should now be scraped up and brushed out. If you are able to acquire an old plastering trowel, you will find it ideal for scraping wooden floors. All rubbish should be dumped on the site rubbish heap. If only small amounts of rubbish have to be removed, they can be put into an empty plaster bag. Do not over-fill bags with wet material or they will split open when you pick them up.

You may find it necessary to go back into the room later in the day, when the floor has dried out, and brush the floor a second time, so that it can be left in a clean condition.

11
FLOOR SCREEDS

PREPARATION
OF THE
CONCRETE BASE

The base is referred to as the oversite concrete.

This must be clean, and free of all plaster droppings. Using a stiff bristle broom, sweep the floor thoroughly. Dispose of any rubbish, and to make sure that no dust remains on the floor, water can be applied and brushed over the whole floor area. You will find that surplus water will be absorbed by the concrete. It is common practice to apply a bitumastic membrane to oversite concrete, to prevent rising damp.

Bitumastic solutions come in metal containers, and care must be taken when removing the lids. Most lids tend to be difficult to remove, but you must be patient. Avoid splashing walls, for any stains on plaster will be difficult to remove, and will stain through the decorations. Should splashes occur, you should immediately wash them off with clean water, using your water brush. The membrane is best applied with a soft broom. Before using the broom, immerse it in water. This will prevent the bitumastic solution from clogging up the bristles. Gently tip the solution from the container. It will be best to work around the floor edges first, working carefully up to the wall. When this is done, start covering the remaining floor area, working towards the door. Two, and sometimes three, coats of membrane are required. More than one coat should not be applied in any one day. Do not walk over freshly applied membrane. Apart from breaking the seal, it will stick to your shoes, be difficult to remove, and risk staining other areas of the building, such as a staircase. If your hands get soiled, wash them immediately. Eye damage caused by bitumastic solutions requires immediate hospital treatment.

After use, wash the brush out thoroughly. The second and third coats should be applied using the same procedure.

Do not dilute or 'thin' bitumastic solutions.

When working on first-floor screeds there will be no risk of rising damp, therefore a membrane will not be required. In this case a cement grout is applied. This is a slurrry mixture of cement and water. Before applying grout, the floor must be brushed and wetted down well. You will find it best to grout only the areas that you will be working on at first. In the case of floors, this will be the perimeter of the room, for you will be laying screeds here first. The remaining area can be grouted as the work proceeds. If grout is allowed to dry out, a good adhesion will not be achieved.

Large floor areas, such as factories or shops, will sometimes take more than one day to complete. In this case, you should contain the preparation of the floor to the area likely to be laid each day. People will most likely be working in other areas of the building, and this will doubtless result in them, at some time, walking in your working area, so preparation should be delayed until you are ready to lay the floor. Where there is no risk of rising damp, PVA (Poly Vinyl Acetate) solutions can be used as a bonding agent. These may be diluted, and you will find details of the manufacturers' recommendations on the sides of the containers. There are many different manufacturers producing this very useful adhesive/sealer. All their recommendations may not be the same, so be sure to check before use.

In some cases you will have to lay a screed on a badly contaminated base such as a garage floor. This will require a great deal of preparation, particularly if oil is present on the base. A diluted household detergent should be applied and the floor well brushed with a stiff broom. You may need to repeat this operation several times to ensure that all the oil is removed. Wash the floor off thoroughly and apply the bonding agent, bitumastic membrane or grout.

Occasionally, no base preparation is required. This occurs when a screed has to be laid on an insulation material such as polystyrene slab or fibreglass quilt. All preparation treatments to floors will need to be protected when the material is brought into the room with a barrow. The best way to do this is to lay down the previously described board runs, two boards wide. At the point where the material is tipped from the barrow, lay down a piece of timber to prevent the metal 'nose' of the barrow from causing damage. Try and tip the material as close as possible to where it will be needed to avoid shovelling material around unnecessarily.

Radiator supply pipes are very often set into the floor screed. You must take great care not to damage them, or a water leak could be the result, and this will only be detected at a later stage when the water supply is connected to the system. By this time the floor will be hard, and the damage caused will result in a great deal of aggravation. The pipes that are encased in a floor screed will have a 'lagging' or insulation material around them to prevent heat loss, which will also cause cracking and possible loosening of the floor screed if it is exposed to heat.

Under no circumstances must this lagging be removed.

When all preparation work is done, you can equip yourself with the tools you are likely to require.

You will need the following:

> Flooring trowel
> Large float
> Level
> Gauging trowel
> Long and short straight edges
> Shovel
> Tile or timber for levelling dots

It is possible to use your scratch float for flooring, provided you either remove, or tap in, the nails first. The gauging trowel may be needed for finishing the work around pipes. The straight edges should be of a manageable length, suitable for the size of the room.

The mix or 'gauge' for floor screeds for domestic dwellings is normally 3 parts of sharp sand to 1 part of cement.

If you are using washed flooring grit, a gauge of 4 to 1 is often acceptable, for grit does seem to produce a more hard-wearing surface. To test the consistency of the mix, you can squeeze some material in your hand. When the pressure is released, the mix should have formed into a compact 'ball'. If the mix falls apart or oozes water, it is unsuitable and should not be used.

SETTING UP
LEVELLING DOTS

The first dot should be set in at the entrance to the room at the base of the door lining. The lining should have been fixed in such a way that the required floor thickness will finish below the lining, and take into account the floor covering thickness that is to follow — either tiles or carpet. Thicknesses of domestic screeds are variable, but are normally about 50 mm maximum.

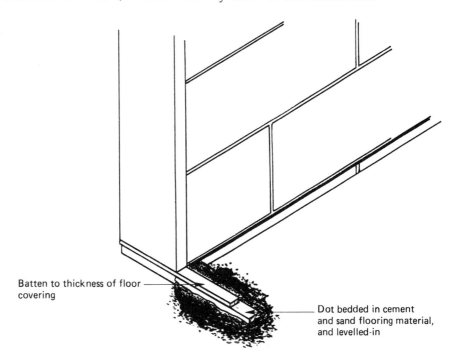

Batten to thickness of floor covering

Dot bedded in cement and sand flooring material, and levelled-in

To insert the first dot, place a shovel full of flooring material on the floor. Spread it out with your trowel, and press in a strip of wood or piece of tile. Set it in, leaving the required distance for floor coverings between the top of the dot and the base of the lining. Place a level on the dot to check that it is levelled-in correctly. Having satisfied yourself that the dot is at the required height, it is advisable to check the thicknesses from this point to the rest of the room using a long rule and level. It could happen that the concrete base is not level. Should the concrete slope upwards away from the door, to a point where pipes and lagging have been laid, it is possible that a level screed would not cover the pipework. In this event the error should be pointed out to someone in authority to discuss possible remedies. No matter what, the floor should be level, and the pipework and lagging adequately covered, so it is likely that any adjustment needed would be at the doorway, and you will need to raise the height of the dot.

This will present problems for the carpenter when he fits the door, and may entail him cutting down the door size. In situations like this there is no easy solution, for someone before you has done a bad job. You must ensure that you do your work correctly, so do not be tempted into laying a sloping screed to get over the problem.

When any adjustments have been made to the dot, you can now carry on with the levelling sequence. The next dot can be placed at a convenient distance from the first, ensuring that your screeding rule can touch both dots. Tap in the dot to the estimated required height, and place your rule on the dots. Now place a level on top of the rule to check that the dots are level with each other. When they are level the bubble in the spirit phial should be central to the two marked lines. If the bubble lies outside the two lines, then that is the dot that is high. If adjustment is required, then the first dot must not be raised or lowered. All correction must be carried out on the second dot. Also check that the dot is level along its own length. Repeat this process around the whole room. Having set in all the dots, the last one can be checked back to the first, to make sure you have not drifted either high or low while working around the room.

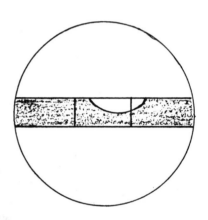

FORMING THE SCREEDS

Lay material between the first two dots and firm-in with your trowel. Rule off the screed with the flooring rule, with each end being in contact with a dot. Take care not to dislodge the dots. If the room is large, it may be advisable to prepare screeds only around a section of the room, for it is not a good idea to allow screeds to lie for too long before working on them, for they will dry out. Make sure the floor/wall junction is cut out cleanly.

LAYING THE MAIN FLOOR AREA

It will be best to start laying at the furthest point from the door, working-in a strip about 300 mm to 450 mm wide between the side screeds. Do not make the strip too wide, for this will cause you to over-reach. This can lead to both poor work and possible injury.

When the work is ruled off, use a side-to-side motion, taking care not to damage or cut into the screeds. Fill in any hollows and repeat the ruling off.

When all holes and hollows are filled, and a good ruled face has been obtained, you can begin rubbing-in with the float, to obtain a good surface.

Lightly trowel up the face of the floor, taking care not to over-trowel or over-polish. Over-polishing will bring cement to the surface, making it liable to cracking. As you proceed with the work, the dots can be removed and set aside for re-use. The holes created by the dots can be filled, and finished off. When your work approaches the area around the door, you will find it necessary to use a shorter rule.

Having completed the floor, the best place to cut off the work will be the centre of the door lining. A wooden threshold is likely to be fitted here, and this will cover the joint between this floor and the next one. You may, however, be continuing straight on with the next floor, in which case, cutting off straight will not be necessary. The two floors will be worked into each other.

The area of floor laid must be protected to prevent animals or other people from walking on it and causing damage.

In hot summer conditions, you must prevent the floor from drying too quickly. This can be done by 'curing'. This means keeping the floor wet for several days after laying. This will allow the cement to set completely. If the floor is allowed to dry out, a complete setting of the cement will not occur, and the floor may break up.

Damp winter conditions will sometimes make curing of floors unnecessary. If a floor is retaining moisture, it will be apparent. You will see 'sweating' on the face of the work. When curing is required, but you are leaving to start another contract, you should inform a responsible person of the fact, and stress the importance of curing the floor. He can then arrange for this to be done for you.

If you do not take this action, and the floor does start to break up at a later stage, the chances are that you will be the person to shoulder the blame.

Curing is done by pouring water on the screed and sweeping it over the whole area with a soft broom.

SCREEDING WITH BATTENS

This method of working is quite acceptable, but probably not as efficient as working with dots. The batten method requires that you lay in battens around the wall edges, level them in, and work from them as you would with soft screeds.

The battens have to be removed, and the channels filled in, and I feel this is both awkward and unnecessary. The only time that I would consider using this method is when the battens are required to be left in the floor to fix carpet edges to. In this event, the battens have splayed sides, being wider across the base than they are at the top. This prevents the battens from lifting, and provides a good fixing for carpet gripper-rods.

12
ROOF SCREEDS

Screeding to roof areas has one distinct difference from floor screeding.

Whereas floor screeds must be laid level, roof screeds will, in almost every case, need to be laid with a slope. The reason for this will no doubt be obvious to you. It is to remove rain water from the roof area and into gutters or drainage points.

Dispersing the water into a gutter at the roof edge will mean that the screed at the edge of the roof will be lower than the screed at the back of the roof.

Alternatively, there may be a central drainage point. In this case the screed at the centre of the roof will be lower than the screed around the perimeter (sides of the roof). In either case, the laying of sloping screeds is called 'laying to falls'.

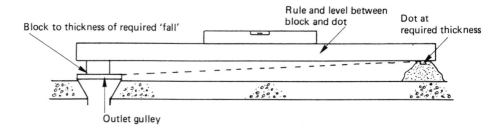

To reduce cost, weight, and to increase insulation, the roof screed can be laid as two separate operations. The setting up of the falls can be done with a lightweight material called vermiculite, mixed with cement. Vermiculite is a larger grade of the fine aggregate used in lightweight plasters, such as Bonding and Browning.

The gauge for a vermiculite and cement screed can vary from 4 to 1, to 8 to 1. You will be told the required mix by your employer.

MIXING VERMICULITE

Mixing vermiculite screeding material can be troublesome because of its light 'flyaway' texture. Vermiculite will be delivered in hessian sacks. These are large, but very light, and easily carried onto a roof for mixing. Mixing in windy conditions can prove difficult as the material, when dry, will tend to blow about.

The cement and vermiculite should first be dry-mixed with a shovel until a uniform colouring is obtained. Water can be added by spraying with a hose pipe. If water is poured into the dry mixed heap it will simply float away. When the mixing is done and grouting of the concrete roof is completed, dots can be set up. This will not be any more difficult than setting up level dots, but it will be necessary to use a different system.

SETTING UP DOTS Assuming the fall from the highest point to the lowest point is 75 mm, the following procedure should be followed.

Set up the lowest dot at the required thickness. Place a block of wood on top of this dot 75 mm thick (for this is the rate of fall required in this example), and set up a second dot at the highest point of the screed, which will be 75 mm thicker than the first. Place a levelling rule and a level between the two dots. One end will be on the top of the block of wood on the first dot. The other end will be directly on the face of the second, higher dot. When a level reading shows on the spirit level, the required fall has been obtained.

Place further dots as required using the same method.

FORMING SCREEDS Screeds can now be formed between the dots. Do not compact the vermiculite mixture, or the insulation properties will be lost.

After ruling off, all you will need do is to pass a float or trowel lightly over the surface. Do not attempt to work up a finished face, for this is only a backing coat, and the face should remain an open texture to ensure good adhesion with the topping-off coat of sand and cement.

PROTECTION AGAINST RAIN When the whole roof area has been completed, you must consider the possibility of rainfall. Should this be likely, the roof area must be covered with a tarpaulin or other suitable material. If rain does fall onto the roof before the cement has set, it will wash away the cement content from the mix and make it useless.

TOPPING-OFF COAT The topping-off coat of sand and cement is best laid the following day to ensure good adhesion.

A mix of 4 parts of sand to 1 part of cement is recommended. The sand and cement coat should be about 25 mm in thickness. Because of the nature of the newly laid base, which offers good drainage, the topping-off coat can be laid a little wetter than ordinary floor screed. Use of the rule, float and trowel will be the same as floor screeding. To establish an even coverage, dots 25 mm thick can be set up. These will need no levelling-in, because this has been done with the vermiculite/cement mixture.

Protection from rain must again be considered, and covering may prove difficult and, at times, impossible. You must consider this before starting such work, and try to do it in settled weather conditions. Roof screeds of this type will later receive a covering of waterproof material, such as asphalt.

IMITATION PAVING On occasions you may be requested to make the screed more attractive by marking out, to form imitation paving.

This is time-consuming and requires great care. Marking out must be done as the work proceeds, for you are unlikely to be able to go back later to do this. As each strip of work is completed, you will need to mark out the positions of the imitation slab edges. When these points are established you will need a clean straight-edge, not your screeding rule; place this against two of the marks to form the first joint line. A piece of 'V'-shaped wood, tacked to the base of your float, can be drawn down the side of the rule and the 'V' joint formed. If carried out with care, the result will be most pleasing.

FILLET

Around the roof edges, at wall junctions, you may be asked to form a fillet, to throw water away from the building. A fillet is formed with sand and cement, pressed into the roof/wall junction with a gauging trowel. This is then rubbed in with a float at an angle of about 45°. A trowel can then be lightly applied to the face of the fillet to give a smooth finish.

WATERPROOFING ADDITIVES

Sand and cement work to roof areas that will not be receiving an asphalt finish will either have a waterproof solution added to the mix, or receive a waterproof dressing after the work has hardened. This will be a clear solution, applied in the manner described for curing floor screeds.

13
EXTERNAL WORK

SCAFFOLDING Work on the outside of building will, at some time, need to be carried out from a scaffold. On houses, you can expect to find a scaffold has already been erected for you, by the builder. *Do not* take it for granted that the scaffold has been left in a *safe condition*.

To check that it is safe, you will first of all need to erect a ladder. Lean the ladder against the scaffold at an angle of about 60°. Do not have a ladder at a steep angle. Before the ladder is used, make sure that it stands firmly on the ground, then to prevent the ladder striking out it must be secured at its base. This can be done by one of two methods:

1. Place *concrete blocks* at the base of the ladder to the height of the first rung. Lightweight blocks are not suitable for this. A short-board ramp can also be incorporated from the ground up to the top of the block.
2. The other method is with pegs driven into the ground against the ladder and then the ladder base can be tied to the pegs.

Having secured the base of the ladder, you must now secure the top. Make sure this is tied firmly to the scaffold. Also make sure that the ladder goes above the walking level of the top of the scaffold so that you have something to hold onto when you reach the top. About 1.5 metres is sufficient. Now your ladder is ready, you can inspect the scaffold.

You should look for *traps* at board ends. Make sure that there are guard rails and toe boards in position.

On lower-level buildings you may have to erect your own sectional tower scaffold. First of all, lay scaffold boards on the ground so that the legs of the tower can stand on them. If you do not do this, and the ground is soft, your scaffold will sink into the ground, and could topple over. Always use the cross-brace sections provided. The tower will be dangerous without them. Always erect a guard rail.

Do not place a ladder against a sectional tower scaffold, for if you do the tower will be subject to a sideways thrusting motion when the ladder is used, and could cause the tower to lean over or topple. It is better in this event to place the ladder against a wall beside the tower.

PROTECTION No matter what type of external finish you are required to do, it will be necessary to protect the surrounding brickwork. This is particularly important when using Tyrolean finishes. It will be essential to 'mask-up' all brickwork and doors and windows. You will find that off-cuts of plasterboard can be tacked onto the edges of the brickwork, and when removed on completion, a nice straight edge will be left to your work, and the brickwork will remain clean.

Health and Safety Executive **Summary sheet for small contractors**

TOWER SCAFFOLDS

Many of the deaths which occur when tower scaffolds are used happen when a ladder is put on the top platform to extend the height of the tower. The tower can become very unstable and if a person climbs the ladder his weight may push the tower outwards and make it overturn.

ERECTING THE TOWER

There are a number of different types of prefabricated towers available, the manufacturer should provide an adequate instruction manual or erection guide for his particular type. The supplier or hirer should pass this on to the eventual erector/user of the tower, who should ensure that it is available on site and that its instructions are closely followed.

Aluminium alloy towers are very light. Make sure that they remain stable and cannot overturn during use or be blown over when left unattended.

The manufacturer's instructions should tell you either the maximum height to which the tower should be erected or, for free standing towers, the maximum height to least base ratio (see diagram). These limitations must be followed. Many UK manufacturers of aluminium alloy towers normally recommend a maximum height to least base ratio of 3:1 if the tower is to be used outside.

If information on the maximum height to least base ratio is not available assume a lower ratio of about 2:1.

Where the scaffold is likely to be exposed to strong winds or where the base is too small for the height of platform needed, the tower must be rigidly connected to the structure it is serving.

ACCESS

The platform must have a safe means of access on the narrowest side of the tower. Do not climb the frame unless it has built-in ladder sections with rungs no more than 300 mm apart and the stiles no more than 480 mm apart. If the frame can be used, climb it from the inside. If not, use internal ladders or stairways fixed firmly to the tower. Never climb up the outside – you may make the scaffold overturn.

GUARD RAILS AND TOE BOARDS

Scaffold platforms from which a person could fall more than 2m (6ft 6in) should be fitted with guard rails and toe boards. The guard rails should be between 910 mm (3 ft) and 1150 mm (3 ft 9 in) above the platform.

BASE OF TOWER

Do not erect the tower directly on recently made-up ground, timber spanning excavation etc.

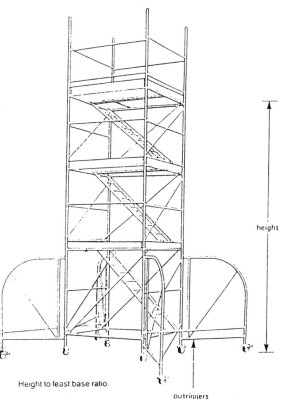

height

Height to least base ratio.

outriggers

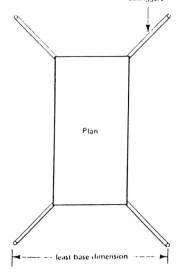

Plan

least base dimension

Tyrolean stains cannot easily be washed off brickwork, and you are also likely to cause damage to your work if you try to do so. Washing off the brickwork may also leave a smeary stain, which will be hard to remove. *Polythene* is probably the best thing to protect windows and doors. This can be taped onto the area to be protected. Alternatively this can be done with brown paper or newspaper. When the job is complete, remove the protective material carefully, so that you do not damage your work. If you find that you have to wash off windows or doors, be careful that water does not splash your work. This is very important when cleaning windows. Too much water applied to the window may run off the sill and onto the finished work below.

PLAIN-FACE RENDERING

For best results, all external work should be carried out in two-coat work, not stronger than 3 to 1 and no weaker than 6 to 1. Try and avoid the use of additives, although frostproofing additives are sometimes required. Some additives have a weakening effect on the mix. Before you start work, you must check the suction of the background. This can be done by applying water to a small area of the wall. If the water is immediately absorbed, there is a high suction. If the water remains visible, or runs down the wall, the suction is low. In either case a light application of water will be advantageous. If you find a mixture of brick and block together, the suctions are likely to differ. You must check this.

During hot summer conditions it is very nice to be in the sun, but more beneficial to carry out the work in the shade by starting on the west-facing wall in the morning and following on with the north, east and then the south walls. This will help to prevent the walls from drying too quickly.

You can now start applying the first coat of rendering. Always start at the top and work down. This first coat should be about 10 mm in thickness, and it will help, when you come to apply the second coat, if you lightly rule off this coat, just to remove any high spots. Having done this, you must key the wall for the next coat. This can be done with a home-made wire or nail scratcher.

At the base of the wall you may be asked to form a drip or bell-cast. This can be done using pre-formed metal sections, or with batten fixed at DPC level. This is sometimes done above window and door openings. If the work goes right down to ground level, it is best to remove some soil from around the building, so that the work goes lower than required. After the job is complete the soil can be returned when the work is hard, to leave a nice neat appearance. It is, however, not a good idea to 'bridge' a damp course, unless a waterproof liquid is used in the mix. The second coat of rendering should not be applied the same day. The mix for the second coat should be the same mix or gauge as the first coat, or weaker. Before applying the second coat, lightly dampen the wall. This is particularly important during hot weather. To prevent the work drying out too quickly, work in pairs when possible — one plasterer applying and ruling off, and the second doing the finishing, or rubbing up, and cleaning off. Holes in the wall left by scaffold putlogs will need attention. Try and drop scaffolds as you go, or get the builder to do so. Ideally, an independent scaffold should have been provided for you, but this is not always the case when bricklayers have used the scaffold during construction.

TYROLEAN FINISH

Tyrolean is applied to two-coat sand and cement work, and the sand and cement finish should be left to the same standard as plain face if a satisfactory finish is to be obtained. The material used for Tyrolean work is Cullamix. It can be obtained in a variety of colours, some of which may need to be specially ordered. Cullamix comes in powder form, and is added to water, and stirred in a bucket. Make sure that the mixture is free of lumps. Always stir the material before taking it from the bucket, for you will find that the heavier grains of Cullamix will sink to the bottom of the bucket if it is allowed to stand for any length of time.

Make sure that you have a suitable, safe scaffold and that all surrounding areas of work are protected before you start.

Tyrolean finish is applied by a hand-held machine. This is of a light plastic or fibreglass construction with a handle at one side. By turning the handle, a centre spindle, which has thin metal blades attached to it, will rotate, and flick the mixture onto the sand and cement surface. It is important always to keep the machine in motion, either from side to side or up and down. If the material is applied with the machine held stationary, a build-up of material will occur on the wall surface, and the mixture will start to 'run' down the wall.

Do not fill the machine up to the centre spindle, for this will cause clogging up, and send out large amounts of material onto the wall. To obtain a good finish it will be necessary to go over the entire work area several times, allowing each application to dry-in slightly before applying the next coat. Clean off on completion as previously described.

PEBBLE DASH

Pebble dash is an ideal finish for use in areas that are exposed to quite severe weather conditions, such as coastal developments. This finish again involves two coats of sand and cement, with small pebbles or crushed stone being thrown into the second coat while it is still wet. As the result will be a rough finish, there will be no float finishing to carry out on the second coat. I think it is preferable to apply the first coat by using the same procedure as for internal floating. That is to say, lay on screeds and rule off, fill in the main areas and rule off again. Then rub in the finished area with a scratch float, but with the nails making slightly deeper scratches than normal.

The second coat must not be applied on the same day. Before applying the second coat, dampen down the first coat, and make sure all glass is protected from the pebble application. The second coat should be a weaker mix than the first, and can be applied at about half the thickness. This coat will not need ruling off, and while it is still wet, the pebbles can be thrown into the surface. Do not allow this coat to dry out before applying the pebbles, or they will bounce off the wall, and not adhere to the mixture.

The method of applying the pebbles is to hold a small bucket of pebbles under one arm, and using a small scoop, or old trowel, throw the pebbles into the wet mixture. Do not stand too close to the wall or you will not get a good 'spread' of pebbles.

When a good even coverage is obtained, lightly tap the pebbles into the mixture to make sure they are all embedded well. This can be done with a trowel, using it very gently. If you have to pick up, and keep, any pebbles that fall to the ground, make sure they are washed before re-use. You will find it helpful to lay polythene under the scaffold to catch the pebbles that fall. This will also make it easier to pick them up.

NEW PRODUCTS
FOR EXTERNAL
WORK

At the time of writing (1989) it is noted that new products are being produced for external work such as polymer-based systems.

Sand and cement rendering failures in recent years can, in some cases, be put down to the use of lightweight blocks for additional insulation. These are of a lower density than traditional blocks and bricks, whereas the density of sand and cement remains the same. This means that modern thermal blocks and sand and cement have different thermal expansion properties, producing a possible cause of render cracking or loss of adhesion.

This will be dealt with in greater depth during your second year's training.

14
MATERIALS

FLOATING-COAT
MATERIAL

Table 14.1 shows some of the pre-mixed bagged materials currently available. Vermiculite or Perlite lightweight aggregates are used in these products. Only the addition of clean water is required.

Table 14.1

Material	Produced by	Suitable backgrounds	Av. thickness (mm)	Av. coverage (m² per tonne)	Setting time (hours)
Carlite Browning	British Gypsum	Common brick walls. Concrete bricks with raked joints. Lightweight blocks. No-fines concrete	11	140	$1\frac{1}{2}$–2
Carlite Browning HSB	British Gypsum	Common brick walls. Lightweight aggregate concrete blocks. Aerated concrete blocks	11	140	$1\frac{1}{2}$–2
Carlite Metal Lathing	British Gypsum	Metal lath	11	65	$1\frac{1}{2}$–2
Carlite Bonding	British Gypsum	Engineering bricks with raked joints. Dense concrete blocks. Stone masonry. Composite ceilings with concrete beams. Cork slabs. Normal ballast concrete. Expanded polystyrene. Gypsum baseboard and lath. Most grey-faced plaster boards. PVA treated surfaces	11	100–165	$1\frac{1}{2}$–2

Thistle Renovating plaster	British Gypsum	Common brick walls. Concrete bricks with raked joints. Engineering bricks with raked joints. Dense concrete blocks. Lightweight concrete blocks. Aerated concrete blocks. Stone masonry. No-fines concrete. For use after insertion of new DPC	11	120	2
Limelite Cement Browning	Tilcon Special Products Division	All solid backgrounds adequately keyed, and of a low to moderate suction, including EML	11	120	12 minimum
Limelite Renovating plaster	Tilcon Special Products Division	Damp walls of older properties, above ground level. All solid backgrounds, after insertion of new DPC	11	130	12 minimum

SETTING-COAT MATERIALS Table 14.2 shows some of the pre-mixed bagged materials currently available. Only the addition of clean water is required.

Table 14.2

Material	Produced by	Suitable backgrounds	Av. thickness (mm)	Av. coverage (m² per tonne)	Setting time (hours)
Carlite Finish	British Gypsum	All Carlite undercoats	2	410–500	$1\frac{1}{2}$–2
Thistle Finish	British Gypsum	Sand/cement undercoats	2	350–450	2
Thistle Board Finish	British Gypsum	Normal ballast concrete. Gypsum baseboard, lath and grey-faced plasterboard backgrounds. Vapour check plasterboards/laminates	3	160–170	1–$1\frac{1}{2}$
Thistle Renovating Finish	British Gypsum	Thistle Renovating undercoat	2	380–420	$1\frac{1}{2}$
Sirapite 'B'	British Gypsum	Sand cement undercoats	3	250–270	$1\frac{1}{2}$
Limelite Finish	Tilcon Special Products Division	Limelite undercoats	1.5	450–535	$1\frac{1}{2}$
Gyproc Veneer Finish	British Gypsum	Plasterboard drylining systems	2	400	1–$1\frac{1}{2}$

Table 14.3 shows some of the pre-mixed bagged materials currently available. Only the addition of clean water is required.

Table 14.3

Material	Produced by	Suitable backgrounds	Av. thickness (mm)	Av. coverage (m^2 per tonne)	Setting time
Thistle Universal	British Gypsum	Ballast concrete. Polystyrene. Plasterboards. Bricks. Aerated concrete blocks. Stone masonry. Metal lath	10 5 13 13 13 13	85–220 depending on background and thickness	2 hours
Snowplast	Blue Circle Industries PLC	As above	10 13	120–130 90–100	1 hour 10 mins

LIMELITE
FINISHING
PLASTER

For use on all Limelite undercoat plasters.
 Benefits include fire resistance and alkaline base, inhibiting mould growth.

LIMELITE
EASY-BOND

Powder for mixing with water for brush application to dense surfaces.
 Mixed and applied as a slurry.

LIMELITE
LIGHTWEIGHT
SCREED

For use on floors and roofs, at thicknesses from 30 mm to 150 mm.
 This material provides both weight-saving and good thermal insulation.
 May be topped with sand and cement screed, or Tilcon LSM ready mixed screed.

PLASTERBOARDS

So great is the choice of plasterboards available that the novice plasterer would be in a state of confusion if any attempt were made to list them all at this time. I will deal only with the most common.

Gyproc Wallboard

These boards are suitable for either ceilings or partition walls.
 One side of the board is grey, for plaster application. The other side is cream, for decoration.
 Available in 9.5 mm or 12.5 mm thicknesses and a variety of sizes.

Gyproc Duplex
Plasterboards

Similar uses as Wallboard, but with one side having metallised polyester film covering. These boards can be either 'Cream and metallised' or 'Grey and metallised' depending on the intended situation for use.
 The purpose of the foil is to provide vapour resistance and improve insulation.
 Again, a variety of sizes is available.

Gyproc Lath

This board differs considerably from most others. It is only 400 mm wide and can be purchased in lengths ranging from 1200 mm to 1372 mm. Thicknesses are 9.5 mm and 12.5 mm. Long edges on these boards are rounded.
 No scrimming is required when plastering, except at ceiling/wall junctions.

Thistle Baseboard

This is another type of board that is smaller than most others, being 900 mm in width and having lengths between 1200 mm and 1372 mm. This board has grey faces to receive plasterwork. Thickness is 9.5 mm. Edges are square.
 It is also available in Duplex form.

BLUE CIRCLE
FLOOR
SCREED SYSTEM

This is a cement-based product pre-mixed with sand and other additives.
 It is used with a pump and mixer combination.
 7   mm thickness is the normal maximum.
 No trowel or float finishing is required.
 This system is self-levelling, and is *not* a traditional screeding system.

15
FIBROUS WORKSHOP

During the second half of your first year's training, much of your time at college will be spent in the fibrous workshop during practical sessions. In this workshop you will be using Plaster of Paris, which is known as Casting Plaster. This type of plaster will set in minutes rather than hours, and because of this, it is essential to keep any tools and equipment regularly cleaned.

Benches that are used in this situation can be topped with a variety of materials, ranging from wood to plaster, marble or melamine. Any surface that is flat and unblemished is suitable.

Assuming that only wooden benches are available, the following steps can be followed to produce a satisfactory bench face. In this case the bench must be well scraped to remove any shellac (a sealing agent) and any tallow (a releasing agent).

After scraping, the bench must be rubbed down with a coarse sandpaper, and then brushed clean.

The next step will be to mix a bowl of Casting Plaster and then spread it thinly over the bench surface.

As the mix starts to steady-up it must be worked into the surface with a flat steel-bladed scraper or large joint rule.

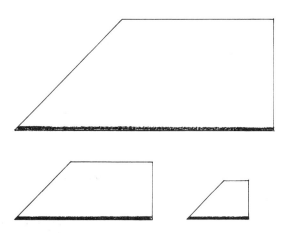

Allow the plaster to harden before proceeding. When the plaster has hardened you will need to rub-in the surface with wet and dry sanding paper. This should be wrapped around a flat block of wood to produce an even surface. Use the sanding paper with water for this operation.

This process will leave a coating of 'fat' on the bench surface. The fat can now be dragged off with a large joint rule to leave a smooth surface.

This surface, as it begins to dry, must be sealed with three thin coats of shellac. Shellac is usually brown in appearance and is available in containers ready for use.

Shake the container well before application, for settlement can occur. For application, a 100 mm wide brush is probably best. Pour the shellac into an open container and carefully apply the first thin coat. Use long even strokes with the brush. Allow this coat to dry-in before applying the second coat. Do not apply a second coat if the first is still sticky to the touch.

Take the same precautions before applying the third coat. After applying the third coat, and when this is dry, the bench should have an even, polished appearance, and is now ready for casting purposes.

This method is only suitable for repair work to existing timber bench surfaces, and must not be mistaken for a plaster top bench.

PLAIN FACED SLABS

The production of plain faced slabs will enable you to appreciate the setting action and working time of Casting plaster.

The size of the slabs you produce will probably be about 600 mm to 700 mm square. Larger slabs will be produced as you become more experienced.

The first stage in the production will be to set up a timber framework. The thickness of the timber framing will determine the thickness of the finished slab. 25 mm is the average. The framing should be of prepared timber. Once the timber is selected and cut to size, the first piece can be secured to the bench with either nails or screws. From this base length, the two side sections can be squared-in and fixed. The fourth section of framing can now be squared-in and fixed.

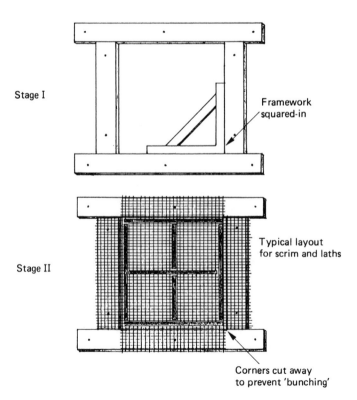

Stage I

Framework
squared-in

Stage II

Typical layout
for scrim and laths

Corners cut away
to prevent 'bunching'

The framing can now be coated with shellac and also the area of bench that is inside the framing.

All materials that will be required can now be prepared. Select and clean two mixing bowls. Cut hessian to size, allowing an overlap of about 75 mm all round. Cut laths to fit within the framework. Laths must be put in soak. The hessian can be soaked and left to drain off.

Hessian is used extensively in fibrous work and acts as a reinforcement in plaster casts.

It has always been my preference to soak hessian prior to insertion into casts, for I feel that a better adhesion with the plaster is obtained and brushing in is made so much easier.

When adopting the practice, it is essential that all water has been drained off.

Soaking hessian is mentioned in the CITB NTI manual, but is not necessarily standard practice. There are (as for many aspects of plastering) regional differences.

The framing and the area of bench within can now be greased with tallow. This allows easy release of the cast, and prevents plaster adhering to the framework.

On completion of all preparation work, put water in both bowls, until they are about half full. The first bowl to which plaster is added is known as the 'firstings'. The second bowl is known as the 'seconds', and to this mix a small amount of retarder must be added. This allows more working time.

Make sure the 'firstings' is well mixed, and contains no dry plaster or lumps. Pour the 'firstings' inside the framework and brush it over the base and up the sides of the framework. Make sure that the plaster is worked right into all corners.

The 'firstings' must be allowed to firm up slightly before applying the 'seconds'. Apply a thin coating of 'seconds' over the base. The prepared hessian can then be laid in, and worked into the plaster with a brush.

The soaked laths can next be laid into the plaster and more plaster brushed over them. The overlapping hessian must now be brought in over the laths and again more plaster brushed in. Make sure the edges of the hessian are clear of the frame.

Central laths can now be inserted and again coated with plaster. At this stage I prefer to put a square of hessian over the whole of the base between the overlapping edges, and brush over with plaster. If needed, more plaster can be mixed for building up the perimeter of the slab, and the central area. Rule off this application with a light rule. Remember that when you are making slabs, what is facing you is the back of the finished slab, and not the face. It is a fibrous slab, not a solid slab, so take care to avoid over-application of plaster.

While waiting for the slab to harden, the surrounding bench and equipment can be cleaned up.

When hard, the slab can be removed. Firstly remove the framework with care, then remove nails or screws and clean off. Store for re-use.

If the bench has been properly prepared, the slab can be easily released by applying pressure from the side.

Turn the slab over, inspect for quality, and carry out any filling or rubbing down as required.

The finished slab can now be used as a base for receiving small panel moulds. These will probably be of your own design, and cutting and fitting to a slab gives you good practice at working-in mitres.

PANEL MOULDS – DESIGN AND PRODUCTION

Before a running mould is constructed to produce panel moulding, you wil need to draw, full size, the shape of the mould you wish to produce.

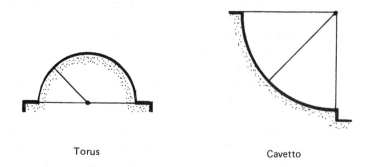

Torus   Cavetto

Your acquired knowledge of geometry will now be needed. A panel mould, generally speaking, should be delicate in its design and normally no more thar about 75 mm in width. I feel that you should try to include some curved features in your first attempt and suggest that you include two of the basic Roman mouldings (others will be included later). The highest point of the mould can be formed centrally with the *torus* with a radius of, say, 10 mm. From its base line include a 5 mm fillet (the flat section) on either side leading into a *cavetto* mould down from both fillets, again with a 10 mm radius. From the base of the cavetto form a further 5 mm fillet either side down to the base of the mould. See illustration.

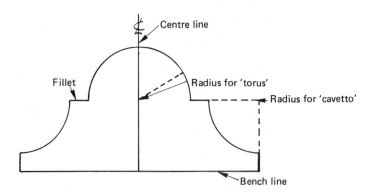

When a satisfactory drawing has been produced it will be necessary to transfer the shape onto zinc, prior to the construction of the running mould, for it is the zinc profile that will form the required mould.

There are two methods that can be used.

1. To transfer the shape, cut out a piece of zinc large enough comfortably to receive the moulded shape. Lay the zinc on the work bench with a sheet of carbon paper on top. Place your drawing on the carbon and lightly follow the mould outline around with a pencil. Your drawing must be retained for later use.

The carbon impression that has been made on the zinc will not be perfect in shape, it will merely be a guide for cutting out an approximate shape with tin snips. The zinc must now be placed in a vice and the required shape produced by carefully filing down the zinc. There is a variety of files for this purpose. During this operation you must regularly check the zinc shape against your original drawing.

2. This method requires that you cut out your drawing and paste it onto the zinc. When the paste is dry the zinc can be placed in the vice and the same filing procedure can begin. Alternatively you may take a tracing from your drawing and paste this onto the zinc.

With both methods you will begin filing with a relatively coarse file, working down to small needle files. When the required shape has been produced, the shaped edge of the zinc can be polished by gently rubbing it with a wire nail. Run your finger along the entire edge length until you are satisfied that there are no irregularities.

The next step will be to 'horse-up' the zinc profile onto the running mould. This means to secure the zinc to a softwood support. The thickness of timber required for a small panel mould need only be about 12–15 mm. For larger moulds, thicker timber will be needed.

Cut off a piece of timber slightly larger than the zinc. Offer the bottom edge of the zinc onto the bottom edge of the timber. Trace around the profile, thus leaving the shape of the mould marked on the timber in pencil. Remove the zinc from the timber and, moving your pencil up from the base of the mould, draw another outline of the mould 3–4 mm away from the original outline.

Place the timber in the vice and cut out the waste area with a coping saw. This type of saw has a fine blade that can be turned at various angles as the cutting proceeds.

What you have now produced is the stock of the running mould onto which the zinc will be mounted. This in turn will be attached to the horse (the section that runs along the bench rule to produce the panel mould).

There are regional variations in the names given to the various parts of a running mould, and those given are common to my own south of England area. As the zinc will only be fixed with small nails, it is advisable to puncture the nail positions in the zinc with a larger nail first. Attaching the stock to the horse can be made more secure if a channel is cut out of the horse into which the stock can be inserted and nailed.

Having made sure that the stock is at 90° to the horse, a brace can now be fitted across the top of the running mould. This also serves as a 'handle' when running of the mould begins. Zinc 'slippers' can be fitted to the horse and stock to allow the running mould to slide more easily (see drawing). When construction of the running mould is complete, it should be coated with shellac. This will prevent moisture being absorbed, thereby preventing any possible swelling of the timber.

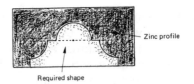

Required shape

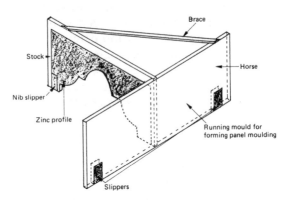

While the shellac is drying, a running rule can be fixed along the edge of the bench. Nails or screws used for fixing should be flush with the top of the running rule. If they protrude above the rule there is a risk of injury to your hands when production starts. The area of bench on which the mould is to be run can now be greased with tallow, to ensure easy release of the mould on completion.

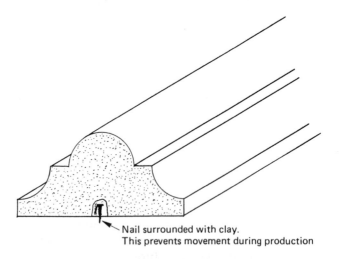

Nail surrounded with clay.
This prevents movement during production

Place the running mould on the bench and mark off the central point of the mould. To prevent the mould from slipping during production, nails are driven into the bench surface along the centre line. They must not hinder the passage of the running mould. To allow for easy release of the completed mould, clay must be placed around each nail.

Select two clean bowls and half fill each with clean water. Add casting plaster to both until the powder reaches water level. Mix the first bowl, but leave the second to soak. Setting of the plaster in the second bowl will be delayed until the mixture is disturbed. The position of the nails in the bench will indicate the central point of the mould. The firstings can now be poured along this line, and the running mould run through. (A bucket of water on the bench is useful for keeping the running mould clean.)

Now reinforcing hessian can be inserted as the work proceeds. It is not likely that laths will be required for panel moulds. As the work progresses, the amount of plaster required will be less at each mixing and the mixture will become progressively thinner.

When the required shape has been produced, you should begin cleaning all equipment that has been used. Leave cleaning of the bench until last so that the mould does not get damaged. When the mould is hard it can be released from the bench. Cut off the rough ends with a saw and insert a small tool between the mould and the bench surface. Run the small tool along the entire length of the mould and it should become free. Lay the completed work on a flat surface — *do not stand it on end*.

PANEL MOULDS - MITRES AND FIXING

When sufficient moulding has been produced, preparations can be made for fixing the work to your plain faced slab. I feel that this is a very worth while exercise for students. By setting up panel moulds around the perimeter of the slab, you will have four internal and four external mitres to work-in. The first step will be to cut the panel moulding you have produced into four pieces, each being longer than your slab by 100 mm. Place one length of moulding along one edge of the slab flush with its edge.

Where the inside line of the moulding rests, draw a pencil line on the slab along its whole length. Repeat this along the other three sides.

Mitre lines can now be drawn on the slab from the corners of the slab to the intersections of the pencil lines you have drawn. All of these lines will be used for positioning the panel mould when fixing takes place. Using a mitre box the 45° mitres can now be cut. Do not attempt to cut the mitres tight as you would with moulded timber. When all four lengths have been mitred, offer them onto

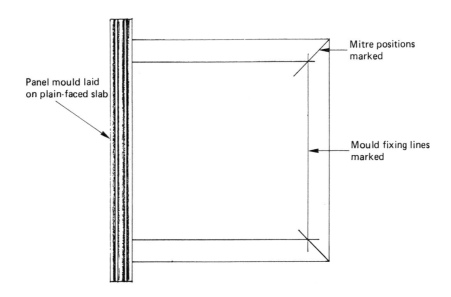

Panel mould laid on plain-faced slab

Mitre positions marked

Mould fixing lines marked

the slab and check that all mitres meet correctly. Make any necessary adjust ments. About a 2 mm gap along the mitres is sufficient.

PVA can now be applied to the back of the moulded lengths and to the slab around its perimeter, between the pencil lines and the slab edges.

Mix up a very small amount of casting plaster and apply a ribbon of the mixture to the slab, centrally between the slab edges and the pencil lines.

Lightly press the length of mould into the wet plaster using the pencil lines as positioning guides. Repeat this process for the other three sides. It is critical that the bed thickness is the same on all sides. Check that the mould level is the same by placing a joint rule across the top of one length and onto the adjoining length.

When all fixing is complete, clean off any surplus plaster that has oozed out from beneath the mould. Filling the mitres can now begin. Do not ever apply wet plaster to dry moulds when mitring. Wet the mitre area.

For the purpose of this exercise, work on only one mitre at a time. For such a small operation, neat casting plaster can be used. Other mixes will be used on larger work. Run some plaster into the mitre groove and allow it to steady up. The filling should be slightly higher than the mould lines. Working-in the mitre will require a small joint rule. In effect it will be used like a feather edge, holding it against the line of the moulded lengths and feathering out the mitre line.

Patience and care will be required for this operation.

REVERSE MOULDS

Reverse moulds provide a means of producing fibrous plaster cornices on the bench. As with panel moulds, the first stage will be to produce a full size working drawing.

This exercise provides an opportunity to use three more basic Roman mouldings — *cyma recta*, *cyma reversa* and *ovolo*.

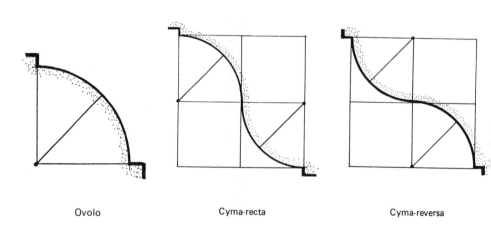

Ovolo   Cyma-recta   Cyma-reversa

The example shown includes each of these and the method of setting out. As a whole they form a *compound moulding*.

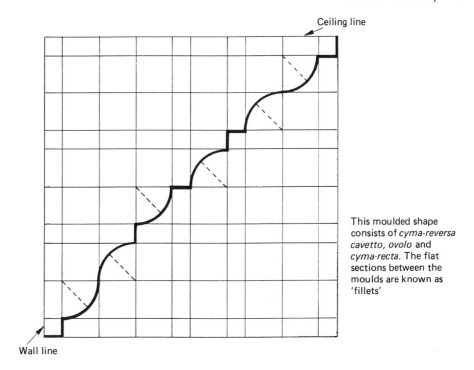

Ceiling line

This moulded shape consists of *cyma-reversa cavetto, ovolo* and *cyma-recta*. The flat sections between the moulds are known as 'fillets'

Wall line

The transfer of the full size drawing to the zinc profile will be the same as for panel moulds. The timber used for 'horsing up' the zinc profile will need to be a little more substantial. When the running mould has been constructed it must be held against the running rule in its working position and again, nails must be driven into the bench surface to prevent movement of the mould, and the nails surrounded with clay.

This type of mould will remain on the bench until castings required to be taken from it are completed. Because of this, it is not really necessary to incorporate any reinforcing in the mould. If however the mould is to be kept for further use, hessian and laths should be included.

When the running of the reverse mould is complete, cut the ends off square, and clean up the tools and work area. The reverse mould will need to be treated with three coats of shellac. Allow each coat to dry before applying the next.

FIBROUS CORNICE PRODUCTION The materials required for casting can now be prepared. Hessian, the length of the reverse mould and its width plus 100 mm, can be cut and soaked for a minute and left to drain off. Laths can be cut to the required length. These should be taken from the water trough where they will have been in soak. The reverse mould can now be greased, and 'firstings' and 'seconds' put in soak, the 'seconds' containing a retarder.

Mix up the 'firstings' to a creamy consistency, making sure it is lump free. Pour the mixture along the mould and brush in well until an even coverage is obtained. Allow this application to steady up and clean the strike-off points. Keeping these clean is most important, for they will form the ceiling and wall lines on the finished cast.

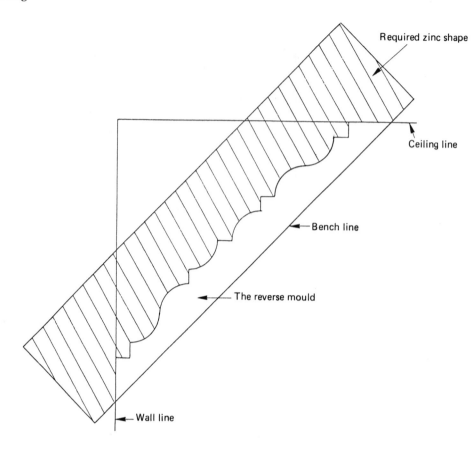

The 'seconds' can be mixed and brushed evenly over the mould. The damp hessian can be laid in the wet plaster. Ensure a 50 mm overlap along each edge. Brush the hessian into the wet plaster. Work the laths into the wet plaster and coat with more 'seconds'. The hessian can now be lapped over the laths and brushed in well.

A lath is *always* positioned at both strike-off points, and on wide moulds additional laths are needed centrally.

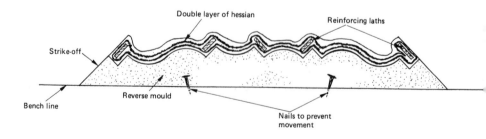

Make sure that you have rather more thickness at the strike-offs, for these areas will be fixing points. For additional strength, small strips of hessian soaked in plaster can be bedded across the mould width at intervals of about 500 mm. Clean strike-offs thoroughly. This can be done with a piece of lath or a joint rule, as shown below. Clean the work area and equipment. The cast will be firm enough to remove from the reverse mould in 15 minutes or so. Gently cut away the surplus material at both ends, ease a small tool between the cast and the reverse at both ends. The two should immediately separate. The cast can be slid along with pressure applied at one end, and removed from the reverse.

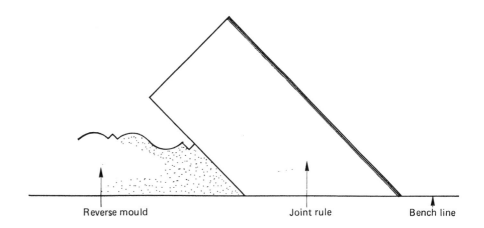

Inspect the fibrous cast and store it flat. The reverse mould can now be cleaned off and re-greased for the next casting.

FIBROUS CORNICES — FIXING

Having produced your own fibrous cornice work, the next logical step will be to gain experience in fixing. This operation requires great care. You will find, at times, that you will have to fit cornice work which you have not made yourself, but which has been produced by a specialist supplier and purchased by your employer.

In this event he will have only purchased enough for the job in hand, and will not expect to have to buy more because of any mistakes made by you. In your college workshop situation, the size of the cornice produced will obviously vary between different students, therefore I will list in detail the suitable methods of fixing both large and small cornices. The basic setting-out principles are always the same. Whether you are fixing cornice made by yourself or by a specialist, you will have a detailed drawing of the cornice, either your own or an architect's. What such a drawing will tell you, is how far the cornice projects onto the ceiling, and how far it drops down the wall. This information is essential. Assume that the 'drop' from ceiling to bottom member of the cornice is 150 mm. Begin on a straight length of wall and make a mark in pencil on the wall 150 mm down from the ceiling, at both ends of the wall. A line must now be 'snapped' between these two points. (On *very* short lengths, this line can be established with a straight edge and pencil.)

To snap a line means to coat a piece of string with either chalk, charcoal or cement powder. One person holds the line against the mark at one end of the wall. The other person pulls the line *very tight* and holds it against the second mark. The line is pulled away from the face of the wall and released. The result will be that a perfectly straight line has been marked on the wall. This will be the line of the bottom member of the cornice.

The same operation can now be carried out to establish the ceiling position of the cornice. Having established both of these lines, you will need to check both ceiling and wall for straightness at these critical points. In your own workshop situation you should have satisfied yourself during the plastering operation that these lines were correct. Even so, they must be checked. In a site situation, the plastering work may have been carried out by someone else who did not take so much care. Any irregular areas must be corrected before fixing begins. When fixing the first length of cornice, there is bound to be at least one internal mitre to cut, to fit the internal angle junction to the adjoining wall.

For cutting mitres, it is advisable to have a mitre box made up specifically for the size of cornice you are cutting. Where more than one length of cornice is needed on one length of wall, it is best to cut the cornice so that the centre joint consists of one internal mitre adjoining one external mitre. For your first attempt however, a straight cut junction will be easier. If the cornice you have produced is small, that is to say less than 150 mm depth and projection, you can use the following fixing method:

(1) Apply PVA to the back of the cornice at what were the strike-off points, which will now be in contact with ceiling and wall surfaces.
(2) Apply PVA to the ceiling and wall along the insides of the 'snapped' lines about 50 mm wide.
(3) Mix a small amount of casting plaster and spread evenly along both strike-offs. Immediately lift the cornice and place it against the two lines and apply a gentle, even pressure. Hold for a short while until the cornice is secure.
(4) Clean off any material that has oozed out from behind the cornice and use it to fill the junction between cornice, wall and ceiling. Wash off with a small soft brush.

For larger cornice work a more thorough procedure must be followed, because it has to be drilled and screwed into position with brass or galvanised screws. On the ceiling, the cornice will be screwed into the joists, and this means that the joist positions have to be located. This can be done by inserting a probe into the ceiling and a bradawl can be used. A bradawl is similar to a small screwdriver with a chisel-like end. Alternatively, an old screwdriver is suitable. Having located the joist positions, place a square against the wall and project the positions beyond the ceiling line and mark with a pencil.

The prepared cornice can now be lifted into position, and the position of the pencil marks on the ceiling transferred to the top edge of the cornice.

There will be no marking required on the bottom of the cornice if it is being fixed to a solid wall, for fixing may be obtained at any desired point. On studwork walls however, the same locating procedure must be followed. The cornice can now be laid flat and holes drilled in the marked positions along the ceiling line. The same must be done at intervals along the wall line. About 500 mm centres should be sufficient. Offer up the cornice again and, with a bradawl, mark the wall positions by piercing through the drilled holes, making a mark on the wall surface. The wall positions that have been marked must now be drilled and plugged with rawl plugs. Fixing can now begin, and for your first attempt, a piece of batten fixed up to, and below, the wall line will be of assistance. The cornice can now be lifted into position and rested on the batten. The screws can be placed along the ceiling line holes and screwed in. The mould will now be supported and more screws can be placed into the pre-drilled holes along the wall line and screwed in.

When the length is complete, the batten can be re-positioned for the next length. Ideally, fixing of fibrous work in this manner should be carried out by two people. When all fixing is complete, mitres, screw holes and edges can be filled and made good. A small tool and a small brush will probably be all that are required for this. A small joint rule is needed on occasions.

Be sure to wet all cornice areas where filling has to take place, and clean off neatly on completion.

This operation assumes that joist fixings will be available at all ceiling/wall junctions. Sometimes the layout of the joists will mean that there is no ceiling joist available. In this event the cornice must be fixed with self-tappers. That is to say, screws that have a thread along the whole length of the shaft up to the head.

When this situation occurs, drill the cornice at any desired point along the ceiling line, and insert a bradawl through the drilled hole and into the surface of the ceiling (not right through the ceiling). By breaking the surface of the ceiling plaster, the self-tappers will screw in more easily.

As with many other aspects of plastering, experience will teach you where time can be saved without sacrificing quality.

16
TOOL AND EQUIPMENT CARE

Good quality tools are never cheap, so it is in your best interest to look after them.

TROWELS

Your *trowels* in particular need constant attention, for they are the items that produce most finished work. Wash your trowel frequently during the working day, and dry it off every night. A little oil applied to the dried trowel will prevent overnight rusting.

Occasionally you will find that the trowel will leave scratch lines on a finished wall. This will mean that your trowel has been knocked in some way and the edge of the blade has become damaged. Inspect the trowel and lightly file off the offending damage. If no file is available, the damage can sometimes be rectified by rubbing the trowel lightly with a piece of clean, soft brick, which should be dampened before use.

A piece of wet brick and a little sand are also a good means of cleaning the back of the trowel blade. Make sure that you do this with the trowel held on a flat base, so that you do not risk cutting your fingers on the edge of the blade.

The trowel handle can also become uncomfortable, either too rough or too shiny. In either case, a light rub with sandpaper will rectify the problem. All other hand tools should receive the same constant attention.

HOP-UP

This is a piece of equipment that is in constant daily use, and it is essential to your safety that if is always kept in good repair.

STRAIGHT EDGES

These must be in perfect condition if they are to produce good straight work, and again need regular attention. Scrape them off at regular intervals. Do not bang them either on the ground or with a hammer. Wash them off every day after use, and lay them down flat.

WHEELBARROWS AND SHOVELS

These should always be kept clean by scraping and washing them off after use. Make sure that wheelbarrow wheels are properly inflated.

CEMENT MIXER

A cement mixer is an essential part of your equipment. Make sure that it is kept not only clean, but also well maintained. This is particularly important with petrol and diesel mixers. Electric mixers possibly need less attention, but after one has been used, make sure the power is disconnected. The working area around the mixer should always be kept tidy. Do not allow the mixer wheels or stand to become buried under a heap of sand and cement.

17
OFF-SITE WORKING

You will find that quite often you will work in occupied houses on either repair work or extensions. Remember this is someone's home, so treat it with respect. Cover all carpets with dust sheets. Do not use plastic or polythene sheets for if either of these materials becomes wet, it will turn into a virtual skating rink.

MIXING WATER

Many homes will have an outside tap from which you can get your mixing or 'gauging' water. Always use it when it is available. Do not put your working bucket into someone's kitchen sink. You will not be thanked by the lady of the house if she finds her enamel or stainless steel sink covered in scratches and the plug hole full of plaster! If you have to take your water from the kitchen sink, then do so with a clean plastic bucket. I have done this, caught the base of the bucket on the sink edge and sent the water all over myself and the kitchen floor, so take care!

ELECTRICAL INSTALLATIONS

When engaged on repair work, you will at some time be asked to make-good the plaster around new electrical installations. Never do this until you have either switched off the power supply, or removed the main fuses affecting the working area. Always remember to switch the power on again, or replace the fuses when your work is complete. The house owner will not want to return from work to find his deep-freeze full of water.

DISPOSAL OF UNWANTED MATERIAL

One problem when working in private houses can be the disposal of rubbish and bucket slops. Try and 'bag-up' your rubbish in an empty plaster or cement bag, so that it can be disposed of easily. Slops are a different problem. Never tip them down the sink or outside drain. Try and find a piece of uncultivated garden. Add plenty of water to the slops, tip them out onto the soil and fork them in, then tip some more clean water on if necessary.

18
FAULTS — LIKELY CAUSES — REMEDIES

DAMP PATCHES ON INTERNAL WALLS

If damp patches appear on the plasterwork on walls of cavity wall construction, the most likely cause will be a build-up of bricklayer's mortar on a wall tie, which is allowing the transfer of water from the outer wall, across the cavity, and through to the plastered wall.

The only way to tackle this problem is to open up the area of wall and clean off the obstruction from the wall tie. This can be done from either inside the building or outside, whichever is the most convenient.

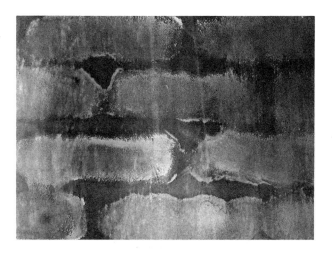

DAMPNESS AT THE BASE OF THE CAVITY WALLS

If this occurs at floor level, or just above, the likely cause is similar to the last item — a build-up of bricklayer's mortar which has accumulated at damp course level, allowing the transference of water from the outer wall. The treatment is as described above — cut out and clear the cavity.

DAMPNESS IN THE CORNERS AT HIGHER LEVELS

This is often accompanied by mould growth, and indicates poor ventilation in the room. Fitting an air brick and plaster or plastic ventilator should rectify this problem. A plaster ventilator for internal use is sometimes referred to as a louvre, and has a fine mesh screen attached to the back, to keep out insects.

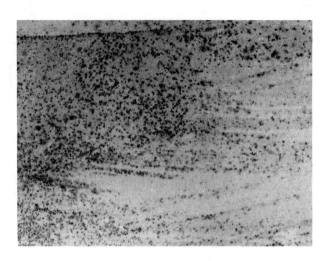

HOLLOW
PLASTERWORK
TO WALLS

This is almost certainly caused by plastering on a poorly prepared background
  This could mean a dusty background, or a high suction background, that ha
not been wet down properly. It could also be a low suction background that ha
not been bonded. The wrong type of plaster may also have been used. In al
cases the hollow work has to be removed, and the background properl
prepared. When this has been done, the area can be re-plastered.

CRACKS IN
PLASTERWORK

These are normally as a result of structural movement. Cut out the crack an
widen it, to allow new plaster to be worked into it. Remove any loose plaste
along the edges. Apply a liquid adhesive to the crack, and work in new plaster
Make sure the plaster is pressed well in.

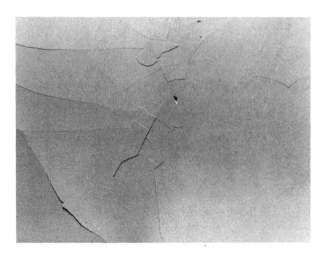

To obtain an almost invisible repair, leave the plaster slightly higher than the surrounding work, and allow it to 'steady up'. When this has occurred, apply water and rub down the joint with a float. This will rub all the work down to the same level and it can then be finished off with a trowel.

BROWN STAINS ON WALLS

This can occur on internal or external work. There are several reasons for this:

1. Dead leaves in the floating material, which indicates poor protection of the sand before use.
2. Cigarette ends allowed to fall into floating material, or thrown into water butts.
3. The presence of clay or soil in the sand used for the work.
4. Poor quality sand, containing iron particles.
5. The presence of metal in the sand and cement work, possibly nails.
6. On older properties, the staining can be caused by the rusting of metal wall ties. Modern wall ties are made of galvanised materials, and this cause is unlikely on modern properties.

In all cases it will be necessary to cut out the affected area to find the cause. Remove the offending matter, and re-plaster.

VERY SMALL CIRCULAR AREAS OF PLASTER FALLING FROM THE PLASTERBOARD CEILINGS

The most common cause of this problem is loose plasterboard nails.

Closer inspection may reveal that a nail has missed the ceiling joist completely. In this case the nail should be drawn out, the area dampened, and made good. Alternatively, the nail may not have been completely driven into the joist. Do this, and make good. This fault can also occur on plasterboard partition walls.

LARGE BULGES ON PLASTERBOARD CEILINGS

Inspection may reveal that the plaster has separated from the plasterboard. Either the wrong plaster has been used or the plaster was applied to a dusty board. Remove the defective area, apply a PVA adhesive, and re-plaster. Take care to leave a good joint between old and new work.

It could be that the plasterboard has come away from the joists. Remove the plaster and check the nails. On bedroom ceilings it is possible that someone may have trodden on the ceiling while in the roof area. Alternatively, a water leak above may have soaked the plasterboard, thereby causing the plasterboard to pull away from the nails because of the soft condition of the saturated board. In either case the defective board must be cut out. Carefully fit a new section and remove enough plaster to be able to scrim the joints between old and new work. Damp down or apply PVA to the edges of the old work.

SEVERE CRACKING OR LOOSENING OF PLASTER ON LATH AND PLASTER CEILINGS

This is quite a common problem, and indicates a loss of the bond between the original plaster and the laths.

The causes can be either damage from above by water or carelessness, or simply the age of the ceiling. This type of repair will take a great deal of time. More often than not, it is advisable to recommend that the whole ceiling be replaced. You often find that you satisfactorily repair a patch, and a few years later the remaining area comes loose. In any case, the banging attached to carrying out a repair may loosen the surrounding area. If you do replace a lath and plaster ceiling it can be tackled in three ways.

1. Remove all the plaster and laths, remove old nails from the joists, and board and set, as you would on new work. (Wear a safety hat, goggles and dust mask when pulling down the old work.)
2. Remove only the plaster from the ceiling, leaving the laths in position. Brush off the laths and ensure that a flat surface is left, with no obstructions. Board and set as for new work, but use longer galvanised nails than normal. Wear protective clothing as described, when pulling old work down.
3. In the case of a ceiling being badly cracked, but still being intact, the following method can be used.

    Locate the joists by inserting a probe into the ceiling. This can be done by firstly tapping the ceiling edges with a hammer. The spaces between the joists will have a hollow sound, but when you tap the ceiling beneath the joist you will hear a dull thud. Having apparently located a joist, the probe can be pushed into the ceiling. An old screwdriver is ideal for this purpose. When you are sure that a joist has been located, mark its position with a pencil line on the wall. Repeat this a little further along the wall. Having located two joists, measure to find out how far apart they are. This will make the job of finding the remaining joists a little easier. When all joists are found, board and set, as with new work. It is advisable to use 50 mm galvanised clout nails when using this method.

WORN FLOOR SCREEDS

If the wear is localised, that is to say, only in small areas, then the repair work required may be minimal. The damage is usually caused by allowing use too soon after laying. You should scrape the worn area to make sure that the floor beneath is sound. If it is, then apply a PVA solution and make good the worn area with a screed levelling compound. This type of material is available from builders' merchants, and has the appearance of cement powder. The only mixing required is to add the powder to clean water. Mix in a bucket as you would do for finishing plasters. The mixture can be poured onto the worn area and trowelled in.

HOLLOW SCREEDS

A screed will only become hollow if the bond between the background and the screed is lost. This will occur if the screed has been laid on a poorly prepared base, or on grout that has been allowed to dry out. All loose areas must be removed, the base properly prepared and the area relaid. Make sure that the area of old floor at its junction with the new work has been well soaked to ensure a good marrying together of the two areas of work.

LOOSE AREAS OF EXTERNAL WORK

This is caused by a lack of adhesion between the top coat and the render coat, or between the rendering and background.

Possible causes are either lack of sufficient key beween the two coats of sand and cement, or poor preparation of the background. Alternatively, frost may have been present in either the background or the render coat when the work was carried out. All loose work must be removed, a PVA solution applied, and either one- or two-coat work replaced, depending on the point at which the adhesion has been lost.

BREAKING UP OF SAND AND CEMENT FACE

The type of finish applied may not have been suitable for the degree of exposure that the work has been subjected to. Weather damage may have occurred. It will be worth considering whether the work should be treated with a sealer/adhesive, and a more suitable finish applied.

It could be that the cement content in the mix was too low. In this case the work should be removed and re-applied in the proper manner. If the defective work can be easily brushed off with the hand, this indicates poor cement content.

CRACKS IN EXTERNAL WORK

When this fault occurs in plain face rendering, much the same procedure can be followed as for internal plaster work cracks, although in this case the final trowel finish will not be required.

Disguising repairs to cracks in pebble dash or tyrolean is very difficult. With pebble-dash repairs you may only be able to cut out the crack and work-in sand and cement. This repair can be disguised by using a stiff hand brush with a stipple action while the material is still soft. The same method can be used for tyrolean repair, using a tyrolean mixture for filling the crack and finishing off with a damp brush.

19
PERSONAL CARE

I think I have now covered most of the areas of work that you are likely to be connected with in the course of your early years of employment. The most important remaining area to be covered is your own personal care and attention. For the building site is not always a comfortable place to work, particularly in the winter, so you must do all you can to make life as pleasant as possible. Try and remember these points:

1. Do not spend your tea breaks sitting on cold and damp concrete. If you must sit on the floor, then put down some dry, empty bags, or you could sit on your material stack, or even your hop-up.
2. Try not to kneel down on a damp or cold floor. Your legs will suffer for it in later years.
3. Purchase some petroleum jelly or vaseline from the chemist's, and apply it to your hands before starting work each day. Just a small amount will be enough, if rubbed into the hands vigorously. Plaster and cement will dry out the natural oils of your skin, so it needs some sort of barrier to help prevent this. If you do not grease your hands, it is likely that sores and cracking of the skin will develop. In severe cases, this may mean that you are unable to work. After work each day, wash and dry your hands, apply a little more vaseline and rub it in well. Then wipe it off with a dry rag. You will find that this application will lift all the dust and grime from the pores of your hands and leave them clean and comfortable.
4. Always wear good shoes or boots. Do not work in wellingtons, for they will make your feet sweat. Leather shoes or boots with steel toe caps are preferable. Clean them off, and when dry, apply polish daily. This will preserve the stitching, and keep them water-tight, and they will last far longer. Trainers or similar footwear may be comfortable, but they will not keep out the dampness that is an inevitable partner to the plastering trade. Nor will trainers give your feet any protection.
5. Make sure your clothing is suitable for your working conditions. In the winter I suggest that a white boiler suit is the best thing to wear. This covers the whole body, and cuts out all draught. Also in the winter, wear more warm clothing than you think you may need. You can always take a jersey off if you get too hot. If you arrive on site in a cold condition, it is not likely that you will get any warmer. In summer conditions, a bib and brace overall is more comfortable. This allows you to remove your shirt if you need, but still gives your body protection.
6. Make sure your day at work is made as comfortable as possible by taking a flask of tea or soup in winter. In summer you may prefer a cold drink. Take enough food for the day. You will not work well on an empty stomach, and a

trip to the cake shop can be expensive. Stay out of the local pub at lunchtimes. It will not improve your ability or your finances.

7. Remember a building site can be a dangerous place, so do not worsen this by sky-larking about, for you risk injury to yourself and others.

I have found these points very beneficial.

20
PRACTICAL CALCULATIONS

THE METRIC SYSTEM

The metric system of measurement is the method now adopted in the construction industry, and is really quite simple.

Unfortunately, you will find on occasions that you will be working with older tradesmen who are reluctant to change old habits. They will still talk in terms of yards, feet and inches. This can lead to some confusion. I do not recommend that you try to use a combination of both systems of measurement, but the following comparisons may be of some help to you in the event of the aforementioned situation arising.

> 25 mm = approximately 1 inch
> 50 mm = approximately 2 inches
> 300 mm = approximately 1 foot
> 900 mm = approximately 1 yard

If, for instance you are asked to get a piece of 4 × 2 timber, the approximate metric equivalent will be 100 mm × 50 mm.

You will find it essential to know that there are 1000 millimetres in one metre. Millimetres and metres are the only terms you will need to learn. Centimetres can be disregarded.

> One metre written in metric form is 1.000 m
> One metre 500 millimetres is 1.500 m
> One metre 250millimetres is 1.250 m
> One metre 750 millimetres is 1.750 m

These figures can be shortened by cancelling out the nought at the end of each example to produce 1.50 m, 1.25 m and 1.75 m. 1.50 m can also be referred to as 1.5 m. Likewise, 1.6 m means 1 metre 600 millimetres and 1.3 m will be 1 metre 300 millimetres.

An inspection of a modern metric measuring tape may be a little confusing to you. You will find that, for instance, 100 mm will be shown on the tape as 10, probably marked in red, and in larger figures. One metre will be marked, probably in red, as 100, and in larger figures. 120 will indicate 1 metre 20 millimetres.

With a little time and study, the metric tape measure will soon become easy to understand.

AREAS

The ability to work out areas of plasterwork is an essential part of your training for the following reasons:

1. Plasterwork is priced by the square metre (m²), and until you are able to calculate the number of square metres, you will not be able to calculate the cost of carrying out plastering work.
2. Until the area of work to be plastered is established, you will be unable to work out the material requirements.

To enable the calculations to be worked out, simple formulas are used.

TO ESTABLISH CEILING OR FLOOR AREAS

The formula is

The length of the room multiplied by the width or breadth of the room

Example I

Length of room 6 metres
Width of room 4 metres

The calculation is 6 m × 4 m = 24 square metres (24 m²).

This is the gross area for either ceilings or floors. If there is an area of work in the room which will not be included, such as a chimney breast, then the dimensions must be deducted from this figure, for this area will not be included either in ceiling or floor work.

For the sake of simplicity, let us assume that the chimney breast is 2 metres wide and 1 metre deep, therefore occupying 2 m × 1 m.

The calculation is 2 m × 1 m = 2 m².

This gives an area of 2 square metres to be deducted from the gross area.

To establish the actual or net area, we must deduct the 2 m² from the gross area.

This will be 24 m² minus 2 m² = 22 m², the actual or net area of work.

Example II

This time, instead of working in complete metre units, we will include parts of a metre.

Again, find ceiling or floor areas, this time with the dimensions of the room being 6.500 m × 4.500 m.

We can break this down to 6.5 m and 4.5 m.

In effect, by ignoring the decimal point, we are multiplying 65 by 45. Therefore, 65 × 45 = 2925.

The position of the decimal point must now be established. The figures that we used, to arrive at 2925, were 6.5 and 4.5.

These two figures have a total of two numbers after the decimal point, and this must be so in the answer. Working from the right-hand side of the answer, count off two figures, and then insert the decimal point, so that 2925 becomes 29.25 m².

No matter how many figures are after the decimal point in the figures used for any calculations, the same total number after the decimal point should appear in the answer.

TO CALCULATE
WALL AREAS

The formula is

the perimeter of the room multiplied by the height of the room

The perimeter means the individual wall lengths added together, and should *include* any door openings. Door and window openings will be added together later, and deducted from the gross area.

Example III

Length of room 5 metres
Width of room 3 metres

If the room is rectangular there will be 2 walls that are 5 metres long, which equals 10 metres, and two walls that are 3 metres long, which equals 6 metres.

10 metres plus 6 metres = 16 m.

We have now established that the perimeter of the room is 16 m.
This must now be multiplied by the height of the room, which we shall say is 3 metres.

$$16 \times 3 = 48 \text{ square metres or } 48 \text{ m}^2$$

There will, naturally be door and window openings in the room which will not be plastered areas. These must be deducted from the 48 m² gross wall area. Let us say that the following items are present in the room

1 door 2 m high × 1 m wide	=	2 m²
1 window 2 m high × 3 m wide	=	6 m²
1 window 1.5 m high × 2 m wide	=	3 m²
Total deductions to be made	=	11 m²

Deduct 11 m² from 48 m² = 37 m² of actual plastered work.

LINEAR
MEASUREMENT

This is the system used for measuring items within the trade that cannot be measured in square metres. With these items, all that can be measured is their length.
Some of these items are

Angles and angle beads
Coving
Cement skirting
Narrow widths of plastering (normally less than 300 mm in width)
Labour to frames, that is to say the labour involved in working up to frames etc.

NUMBERED ITEMS These are normally considered, or allowed for, on domestic developments. On large contracts, where a Bill of Quantities has been prepared, they will be listed.

BILL OF QUANTITIES A *Bill of Quantities* is a list of all work to be done, and a section is compiled for each trade. The types of work likely to have a Bill of Quantities include shops and industrial development. Things that will be listed as numbered items include

> Pipes
> Heating ducts
> Electrical installations

In the case of 25 mm pipes, the item should be written as

> Labour to 25 mm pipe — 15 No.

In other words you will have to plaster around 15, 25 mm pipes.

Once you have learnt how to calculate areas of work, you will have the basic requirement for calculating the amount of materials needed and their costs.

PRINCIPLES OF PRICING Pricing for plastering can be done in two ways:

1. Pricing for only supplying labour, with the materials being supplied by the builder or client. This is known as 'Labour Only'.
2. Pricing for supplying both labour and materials required. This is known as 'Labour and Materials'.

Even if you are pricing labour only, the client may ask you to let him have a list of materials required. As you can see, it is essential for you to have a good knowledge of all areas of practical calculations.

AMOUNT OF MATERIALS REQUIRED To calculate the amount of materials required, you will need to know the covering capacity of different materials. That is to say, how many square metres of work can be covered with a bag of plaster. If you go to your local builders' merchants they will supply you with a *British Gypsum Plaster Selector Guide*. Not only will this advise you on suitable plasters for different backgrounds, but will also list the covering capacities of the different plasters.

The covering capacity will be given as the amount of work that can be covered by one tonne of material. This need not cause any confusion. You will need to know that there are 1000 kilos to 1 tonne. There are 20 bags of plaster to 1 tonne. Therefore each bag weighs 50 kilos ($1000 \div 20 = 50$).

Using the *Plaster Selector Guide*, we can show some examples. Carlite Browning is stated as covering between 130 and 150 m^2 per 1000 kilos. An average figure will be 140 m^2 per 1000 kilos (1 tonne); so if 140 is divided by 20, we can see that one 50-kilo bag will cover 7 m^2 of wall area.

Board Finish is stated as covering between 160 and 170 m² per tonne. Average coverage is therefore 165 m² per tonne, and 165 divided by 20 equals a little over 8 m² per 50-kilo bag.

Sirapite will cover an average of 260 m² per tonne. 260 divided by 20 gives a coverage of 13 m² per 50-kilo bag. With all of these figures the coverage depends on the type of background, and the thickness of plaster applied.

Let us now assume that there are 100 m² of plasterboard to be set with Board Finish. We know that one bag will cover about 8 m², so to find out how much plaster is required to cover 100 m² we simply divide 100 by 8. This works out at 12.5 bags. As half bags cannot normally be purchased, 13 bags will be required.

PRICING — LABOUR ONLY

We can now consider the two types of priced work, labour only and labour and materials.

Labour only is the easiest to work out, for once the area of work is established it has only to be multiplied by your required sum per m². One plasterer may be happy working for £4.00 per square metre, while another may consider his work is worth £5.00 per square metre. Again the figure may be related to the amount of work and the site conditions.

Assuming you are charging £5.00 per m² for 100 m² of plastering, the labour only cost will be £500.00.

PRICING — LABOUR PLUS MATERIALS

Should you also be required to supply materials, you will first of all need to find out the cost of the materials. Let us assume that you are applying Board Finish to 100 m² of plasterboard, as in the earlier example. We know that 13 bags will need to be purchased. For the sake of this exercise we will say that the plaster will cost £3.90 per bag; 13 bags at £3.90 will cost £50.70. This figure will give an approximate material cost of £2.00 per metre. Therefore we have labour only at £5.00 per m² and labour and materials at £7.00 per m². After a time, and if material prices remain stable, you will find that you will be able to compile a list of appropriate prices for different types of work.

You must appreciate that the figures given above are only examples, and are not intended to indicate current prices either for labour or materials.

COVERING CAPACITY OF SAND

Not all plastering materials are delivered in 50-kilo bags, the main exception being sand. You may find that working out the covering capacity of sand is a little more difficult.

Sand is sold either by the tonne or by the cubic metre (m³). There does not, at the time of writing, seem to be any uniformity between various suppliers, but as an approximate guide 1 m³ of sand weighs 2 tonnes. Before sand is mixed with cement and water, it will be of a loose nature and, if spread out will cover a greater area that it will when wet. This is an important fact to remember and the reason will be given shortly.

All calculations for sand requirements will need to be worked out in cubic metres (m³). A cubic metre is a cube 1 m × 1 m at its base, and 1 m high. Of course, sand will not be delivered in this form but, to calculate your requirements, it is necessary to consider it in this way.

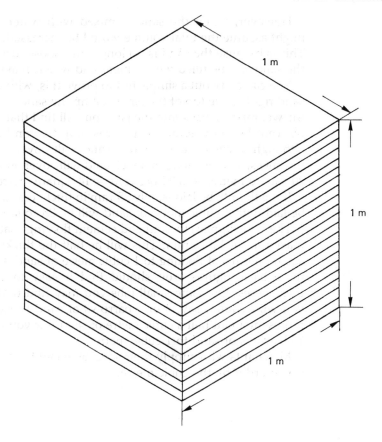

Let us take an example. We will assume that we are laying floor screeds at a thickness of 50 mm. Now we can draw a cube to represent 1 cubic metre. If we divide the cube into layers 50 mm thick, we can see that in the height of the cube we have 20 layers. This tells us that 1 cubic metre of sand, if spread out at 50 mm thickness will cover 20 square metres in area.

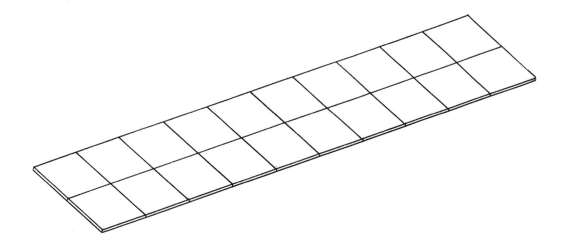

However, when the sand is mixed with water and cement, although you might assume the total volume would be increased, the opposite in fact occurs. This is because the sand is no longer in a loose, dry condition. All the voids in the sand will be filled with cement and water, making the sand more compact.

You can carry out a simple test to prove this, with a jar of dry sand. Fill the dry sand right to the top of the jar. Then tip the sand out and mix it with water. Put the wet mixture back into the jar. You will find that the mixture is now about 30 per cent less in volume than it was in a dry condition. This then tells us that although 1 cubic metre of dry sand will cover 20 square metres at 50 mm thickness, 1 cubic metre of a wet sand will cover about 30 per cent less area. Or 20 square metres less 30 per cent = 14 square metres.

If the floor to be laid is only 25 mm in thickness, then the area of floor that can be laid with 1 cubic metre of sand will be 28 square metres.

The same applies to sand and cement floating coats, which will be at a normal maximum of 12–13 mm. This will be half the thickness again, thus giving 56 square metres of floating to 1 cubic metre of sand. Once you have remembered these basic figures, you should find it easy to calculate for work at other thicknesses. On some small jobs you may find that you can easily assess that you will require, say, 4 or 5 barrows of sand to carry out the work. This will represent about half a cubic metre of sand, for you will find that there are up to 10 barrows of sand to 1 cubic metre.

Merchants who supply their sand by weight will do any conversions for you if you order by the cubic metre.

21
TRADE TERMS

Additives A material or substance added for various reasons. Liquids are added to sand and cement mixes to make them more workable, largely to replace the use of lime. These help to retain the moisture in a mix, and are known as 'plasticisers'. Liquids can be added to sand and cement to either frost-proof or water-proof the work.

Adhesion The bonding together of two coats of material, or material to background.

Angle beads A pre-formed metal section placed at external junctions to produce a knock-resistant angle.

Angles The points where two surfaces meet — wall, floor, ceiling or beams. Where two walls meet in a corner of a room is an internal angle. Where two walls meet, such as the face and sides of a chimney, is an external angle, also referred as an **arris**.

Arris *See* **Angles**.

Background The structural substance to which plaster is applied.

Banker board (or Banker box) A flat base set up to mix rendering, floating or flooring materials on, by hand. Sometimes set up in front of a cement mixer to tip material onto, to keep it in a clean condition. When used for Carlite Browning or Bonding plasters, it has side boards fitted to retain the mix.

Bell cast A gentle curving shape, most common at the base of external rendering at damp-course level, to direct water away from the building.

Bill of Quantities *See* chapter 20.

Bond The efficient working together of two coats of material or the material to a background.

Bonding agents *See* **PVA**.

Catch boards Scaffold boards placed around the base of a wall to catch any floating material that falls to the floor. Particularly useful on concrete floors.

Cove (Gyproc) A pre-formed decorative plasterboard section applied to the ceiling/wall angle. Often used to cover cracks at this point.

Coved angle   A curved finish to an internal angle. Can be any size, and the usual reason for its presence is for easy cleaning and hygiene. Common in hospitals.

Coving trowel   A purpose-made tool for forming coved angles.

Curing   Preventing sand and cement work from drying out too quickly by applying water to the finished work. Not to be carried out within 24 hours of finishing work.

Dabs   Small amounts of plaster used to secure angle beads.

Datum line   A line set up with the use of a water level when working on ceilings and floors.

Devil float (or Scratch float)   A wooden or plastic float with small nails tacked through the ends to provide a key on floating coats.

Dots   Pieces of wood, tile or slate set into the material and levelled in, for setting out work on ceilings, walls and floors.

Dubbing out   Most common on old walls that are in a poor condition. The act of filling in hollow areas.

Expanded Metal Lath   Metal sheeting or metal strip, being diamond-shape patterned. Used on ceilings, partition walls, in sheet form. For covering wall plates in strip form.

Falls   Laying of screeds with a slope. Used on roof work.

Feather-edged rule   A rule with one side tapered to allow working into internal angles when straightening up the work.

Feathering-out an angle   The act of using the feather-edged rule to straighten an angle.

Fibre, Fibrous   Lightweight decorative mouldings.

Gauge   Gauging-up, meaning to provide a mixture of plaster or sand and cement. To knock-up.

Gauge rod   A measuring rod, cut to plasterboard length, to save continuous use of the tape measure when tacking plasterboard.

Gauge rule   A piece of shaped rule used to rule-off reveals to window and door frames, to ensure a parallel margin around the frame.

Green   The condition of sand and cement work before the final set has taken place.

Green suction   The early degree of suction on sand and cement coats. A cool suction, not severe.

Grout   A cement and water slurry. Normally applied to concrete bases to create a bond for the screed.

Key   Scratchmarks on backing coats to ensure a good adhesion. A background material with a 'good key' has a face that will readily accept plasterwork.

Knock-up   *See* **Gauge**.

Lath   A wooden slat, common on old ceilings (lath and plaster). Still used in fibre workshops.

Lining   A door frame of the type used for internal door openings, set into door openings to provide a working line for the plasterer.

Making good   Repair work.

Plasticiser   *See* **Additives**.

Pricking up   The first coat of material applied to expanded metal sheeting on ceilings and walls.

Plumbing   Making sure that a wall is perfectly upright by using a rule and level.

Pug   Sometimes the name given to floating material. A term mostly used by bricklayers.

PVA (Poly Vinyl Acetate)   A liquid adhesive/sealer with many applications. Manufacturer's instructions must be observed. PVA is a bonding agent.

Scratching   Keying one coat of work to receive the next.

Screeds   Strips of material, levelled-in as points from which to work-in main areas of work. Used on ceilings, walls and floors. Areas of cement and sand flooring are also referred to as *floor screeds*.

Scrims   Hessian and cotton are the most common. Used for setting into plaster over plasterboard joints to prevent movement and cracking. Hessian is the stronger, but cotton would appear to be more popular, being less bulky. Hessian is also used in fibrous plastering.

Skins   The inner and outer walls of a building.

Slurry   *See* **Grout**.

Soffit   The underside of a beam, window or door lintel.

Straight-edge   Any type of *prepared* timber that has been straightened along it working edges and suitable for all types of plastering.

Studwork   Mainly timber framework, set up to create partition walls. Suitabl for receiving either plasterboard or EML sheets.

Suction   The rate at which moisture is absorbed into a background or backing coat.

Sweating   The water that is visible on the face of finished work, usually i damp conditions.

Thinners   A liquid used for diluting a substance.

Thinning   The diluting of a substance.

'V'-joint   A hand-worked feature in cement work. Described in chapter 12.

INDEX